MODERN RIFLES SHOTGUNS & PISTOLS

MODERN RIFLES SHOTGUNS & PISTOLS

IAN V HOGG

Exeter Books
NEW YORK
A BISON BOOK

AUTHORS NOTE

In considering the development of firearms today, it is necessary to make several subdivisions and groups. But a fundamental division to be made from the start is to divide the weapons into two classes, *sporting* and *combat*. These can then be further subdivided: sporting, in a firearms context, can mean either competition shooting or hunting. Combat weapons can be sub-divided, since they can be classed as entirely military weapons or as weapons of a combat type, used by police and paramilitary security forces. As we shall see, although these divisions may appear to be quite clear cut, there is a certain amount of overlap between them.

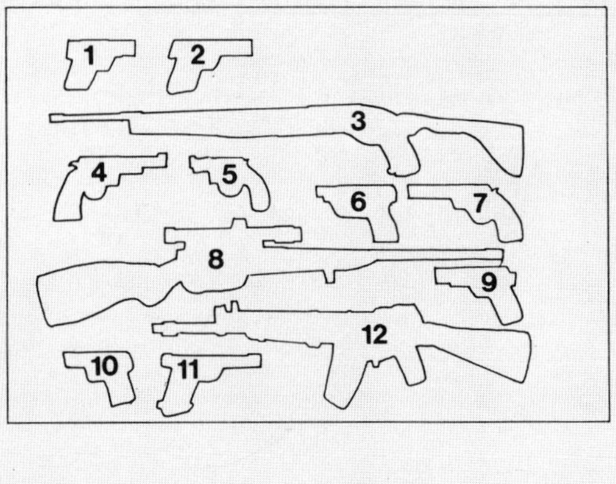

1 Colt Officer's Model .45 ACP
2 Colt Model G70 .45 automatic
3 Smith & Wesson M3000 shotgun
4 S & W M29 .44 Magnum
5 S & W M66 .357 Magnum
6 Heckler & Koch VP 702 9mm auto
7 S & W M686 .357 Magnum
8 Sporter SMLE No 4 in .308
9 Arminex Trifire .45 ACP
10 S & W M469 9mm auto
11 Agner M80 .22LR Target pistol
12 Valmet M76 .223 semi-auto rifle

First published in USA in 1985
by Exeter Books
Distributed by Bookthrift
Exeter is a trademark of Simon & Schuster, Inc.
Bookthrift is a registered trademark of Simon & Schuster, Inc.
New York, New York

Produced by Bison Books Corp.
PO Box 7608
Greenwich CT 06830

ISBN 0-671-07526-8

Printed in Hong Kong

CONTENTS

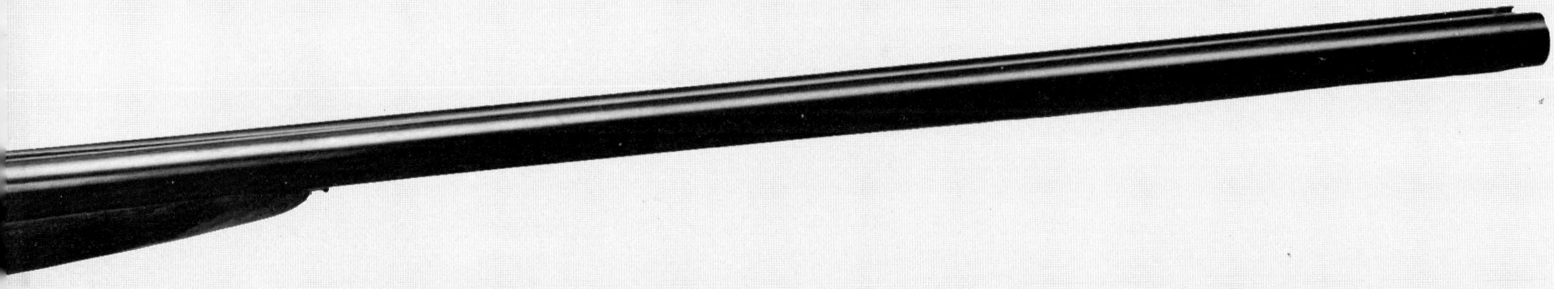

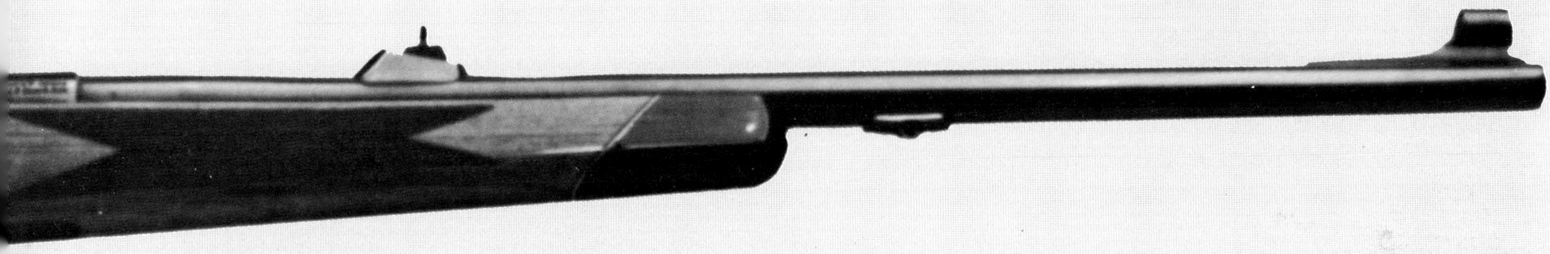

Top: W & C Scott's traditional style Chatsworth model 12-gauge shotgun.
Center: Colt-Sauer Magnum hunting rifle.
Bottom: Colt Diamondback revolver in .22LR caliber.
Following page: Close-up of the action of a fine Winchester side-by-side 12-gauge.

WINCHESTER

SHOTGUNS

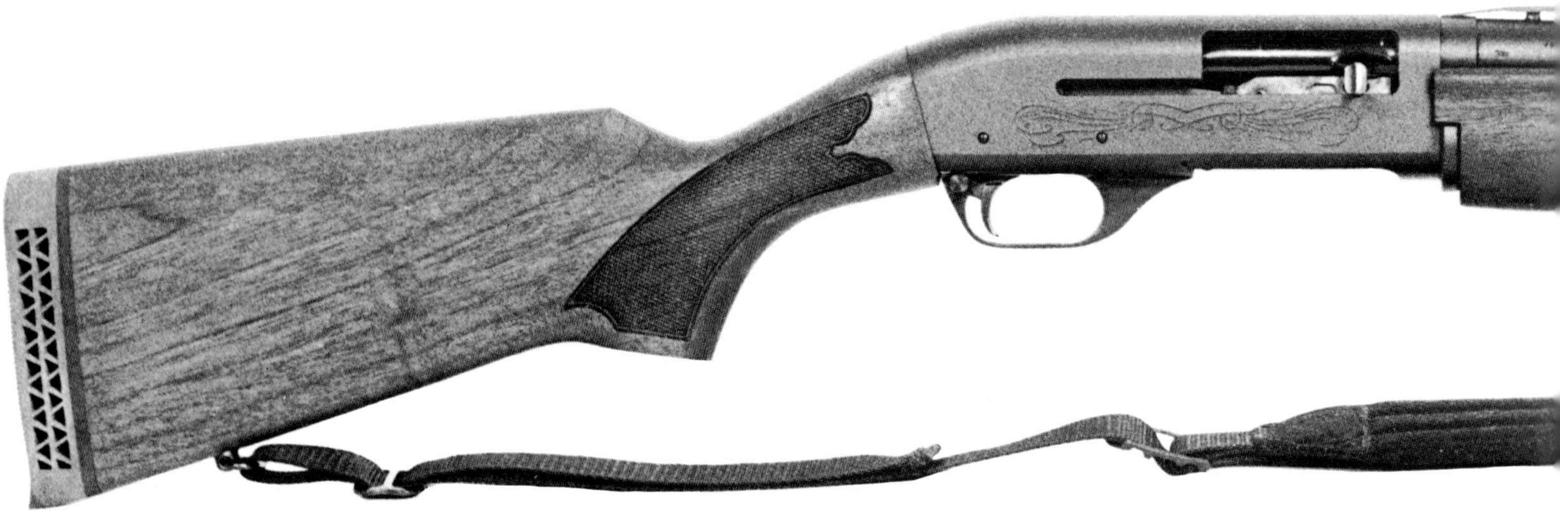

Trying to keep up with shotgun manufacturers is difficult; new names appear and disappear almost before one can register their presence. The old-established makers go on from year to year, but the current upsurge of shotgunning has led to a number of companies coming into a highly competitive market. When they find out just how competitive it is, they soon decide to take up some other aspect of engineering, and only those firms producing a reliable product at a reasonable price survive.

The reason for this activity is simply that shotgunning has boomed over the past two decades; more leisure, more surplus cash, and people who would never have thought of shooting as a sport thirty years ago are now thoroughly involved. Not, however, in game shooting; the number of game shooters varies relatively little, and, particularly in Europe, the land available for game shooting dwindles a little more every year. The rise in shotgun sport is almost entirely due to Trap and Skeet shooting, conducted on commercially-organised ranges in easy reach of towns and cities. In Britain, where shotguns require permits, the balance between the number of permits issued to country-dwellers and town-dwellers has shifted enormously in the past 15 or so years, almost entirely because of this new leisure activity.

For those who tend to confuse Trap and Skeet (as almost everybody does to begin with) perhaps a few words of explanation might be in order. Trap shooting, which is now an approved Olympic sport, has the shooter positioned 15 meters behind a trench which contains a number of machines which will throw targets into the air. The 'Olympic Trench' has 15 machines, the 'Universal Trench' has 5 machines, and the 'International Ball-Trap' trench has but one machine. The machines are adjustable, allowing them to fling the targets in a variety of trajectories within the limits which are specified by international rules. The sequence, speed, angle, height and direction of the clay targets varies with each one. The shooter is allowed two cartridges for each target; five shooters compete at a time and change places in the line after each target launch.

Skeet is fired on a semi-circular layout. At each end of the 40-yard semi-circle is a 'release house' with a target-throwing machine, the house on the left being the 'high tower' and that on the right being the 'low tower'. The shooter progresses to eight firing stations in succession, beginning in front of the High Tower and working his way around to Position 7 in front of the Low Tower and finally Position 8 half-way between the two towers. At each station a target is thrown first from the High Tower and then from the Low; one cartridge is allowed for each target. On stations 1, 2, 3, 4, 5, 6 and 7, after firing at single targets, the shooter is presented with a 'double' – a target thrown from each station simultaneously; the

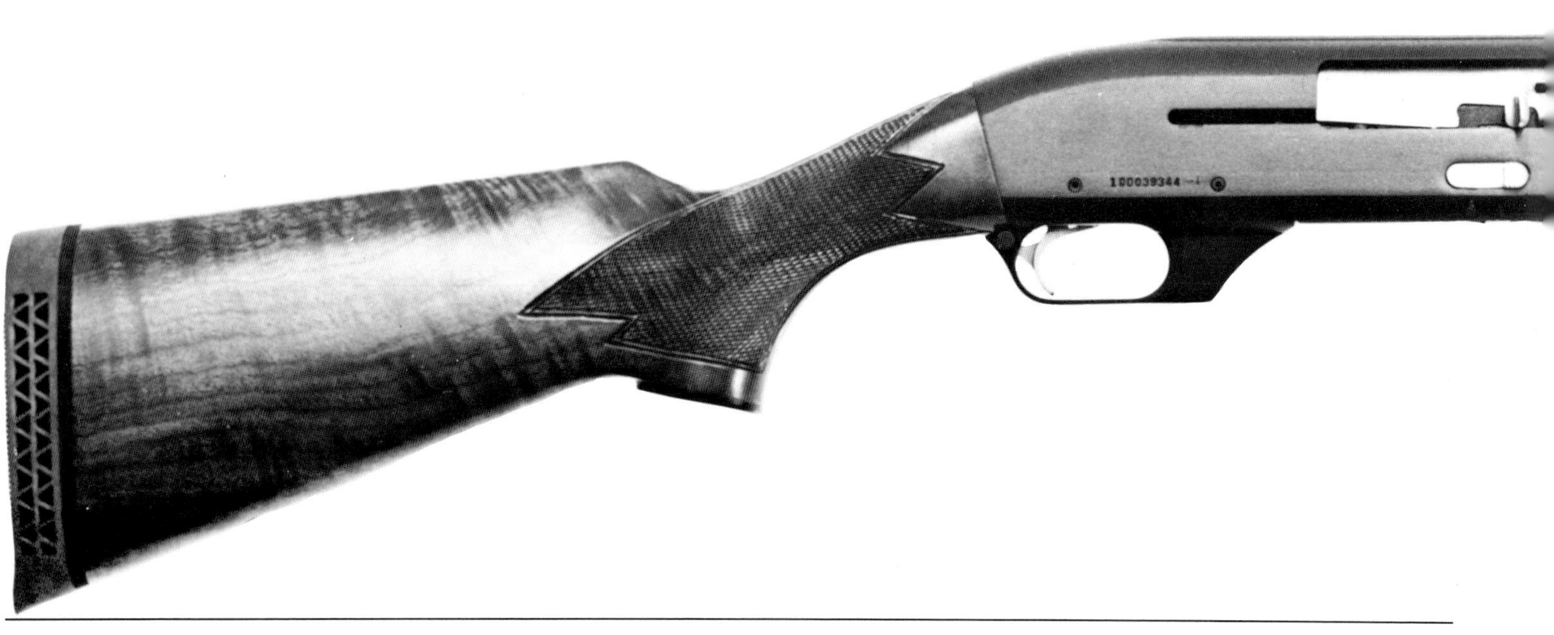

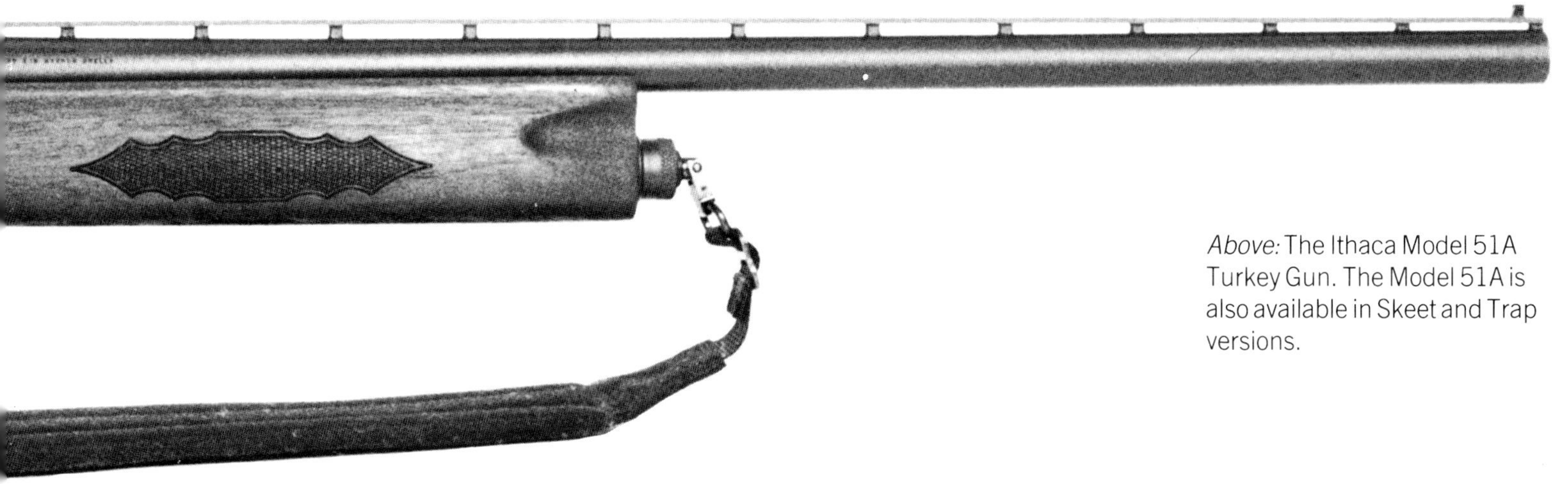

Above: The Ithaca Model 51A Turkey Gun. The Model 51A is also available in Skeet and Trap versions.

shooter has two shots and must take the target from the nearer tower first.

A third type of contest, which originated in Britain, has yet to become widely-known internationally; this is 'Sporting Trap'. Normal skeet and trap tend to have an artificial air about them, and Sporting Trap has been devised in order to introduce a more natural environment. A stretch of countryside with bushes and trees is used, and traps are set in various places to launch at angles which reproduce various types of target, from high-flying pheasants to rabbits scuttling along the ground. The shooter can stand in one place, or he may be able to follow a marked trail, being presented with targets as he walks. As might be imagined, Sporting Trap demands careful planning of the course, with particular attention being paid to the safety aspect, but it can provide a most realistic contest.

The growth of competition shooting of this sort has been reflected in the types of guns reaching the market. The over-and-under double gun is practically compulsory for trap and skeet, one reason being the claim that with a single barrel over which to sight the chance of hitting the small clay target is improved over that of hitting with the traditional side-by-side double gun. But the classic side-by-side survives and is the universal game gun, with a range of choice which covers every ballistic requirement and every depth of pocket; the fabulous guns which are frequently reported in newspapers as fetching prices comparable to a comfortable house are invariably best-quality side-by-sides.

Automatic shotguns come in two kinds; those operated by recoil and those operated by gas. The recoil pattern usually owes something to the original Browning design of 1900, a 'long recoil' system in which barrel and bolt recoil on the stock, locked together. At the end of the movement the bolt is held fast and the barrel propelled forward by a spring, so that the spent case is extracted and ejected, after which the bolt is automatically released to run foward and load a fresh round. This gives a very soft recoil action which most shooters find more comfortable than a normal gun.

The alternative system uses gas tapped from the barrel, behind the shot charge, to drive back a piston running in a tube beneath the barrel. This operates the bolt to extract and reload. There are numerous small variations in design, each claiming to give some advantage which others do not have, but the system is reliable and efficient, though some shooters claim that the recoil is slightly harder than from a comparable long recoil gun.

Another popular system is the slide action, in which the firer's hand pulls back the fore-end and thus mechanically actuates the breech mechanism. Like the slide-action rifle, this has the advantage that the hands never leave their firing

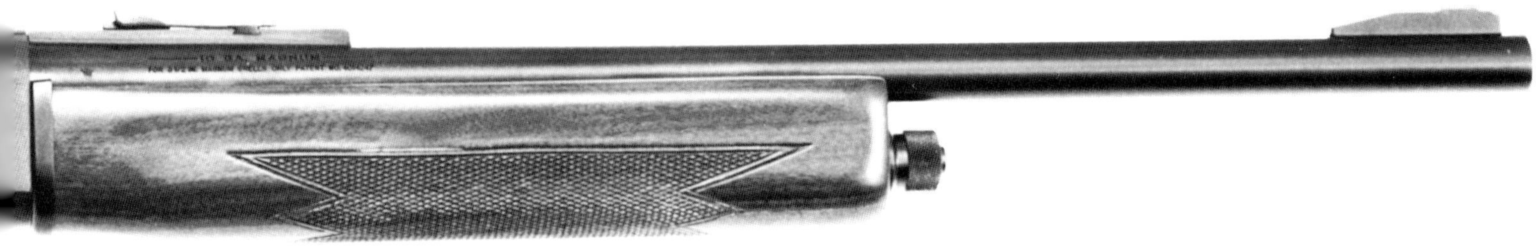

Above: The powerful Ithaca MAG-10 Deerslayer.

A selection of Winchester shotguns and accessories. From top to bottom, Diamond Grade Skeet, Supergrade Skeet and Diamond Grade Sporter.

position and a second shot can be got off very rapidly.

One could ask, at this point, why bother? If a side-by-side or over-and-under has two barrels, then two shots can be loosed off without the need to operate a slide. The answer is that the slide gun (like the automatic) is a single-barrel weapon, and some people prefer that for ease of aiming, and also that the slide gun invariably carries more than two shots in its tubular magazine – anything from 3 to 8 can be found. So that if the second shot misses, a third can be got off very rapidly to try and save the day.

This availability of ammunition in repeating guns, whether slide or automatic, is something which can lead to abuse in the wrong hands; the first day of the hunting season in Spain and Portugal used to sound like the Battle of the Marne being re-fought as hunters blasted shot after shot at anything which moved. (Wise men stayed indoors on that day.) Today, however, most countries have instituted regulations about magazine weapons which, in golfing terms, 'handicap' the hunter and give the game a better chance. Many states in the USA, for example, insist on no more than three shots in the magazine, and shotgun manufacturers provide plugs for larger magazines to limit their capacity.

Viewing the profusion of guns available, the beginner's first problem is generally of what caliber or 'gauge' to select, and here, as in so many other subjects, the answer has to be 'What do you intend to do with the gun?' Is the intention

Below: Belgian police officers with FN-Browning slide-action 12-gauge guns.

Bottom: The mechanism of a Sturm Ruger over-and-under 12-gauge.

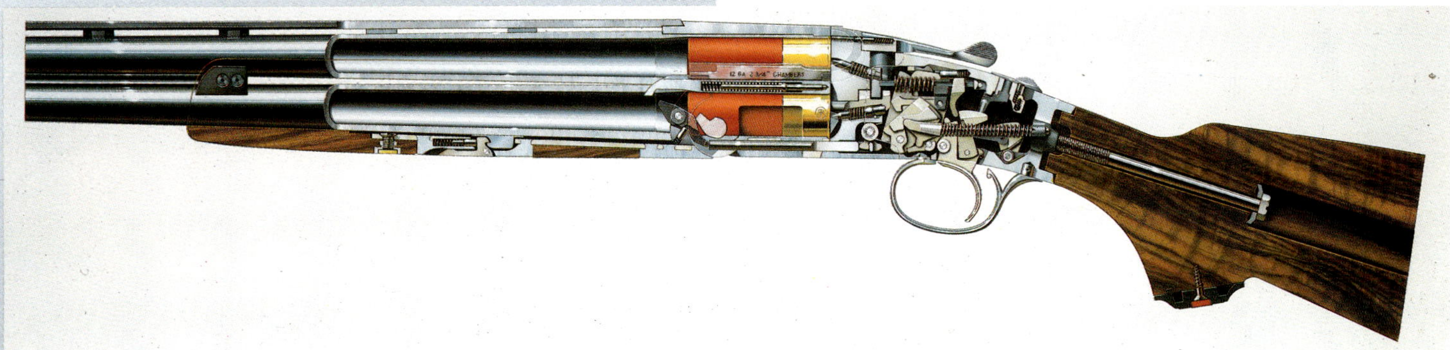

Contrasting views of the breech design of a Ruger over-and-under 12-gauge (below) and of a more traditional design, the Chatsworth, by W & C Scott (below right).

The Ruger gun field stripped.

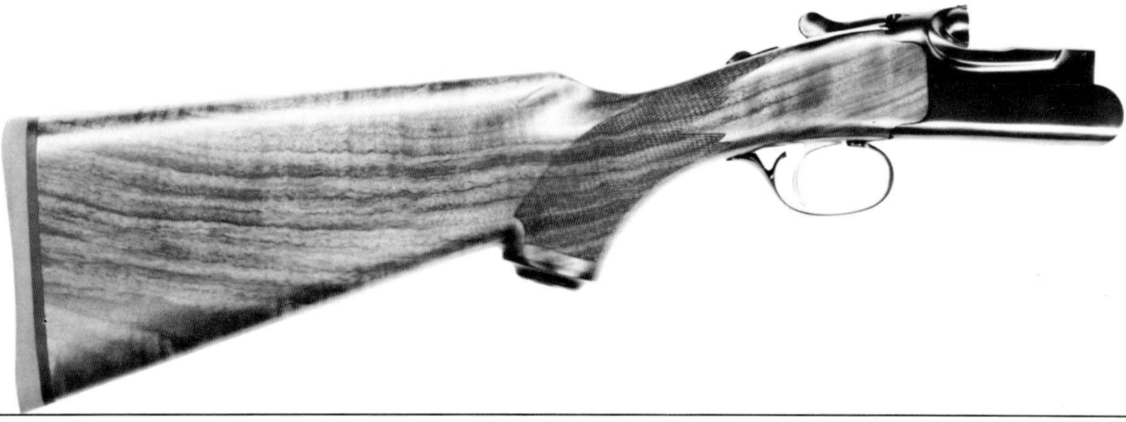

purely competition shooting, or ducks, or grouse, or rabbits or what? The farmer who wants to keep a gun on his tractor on the chance of potting a rabbit for supper will be satisfied with a .410, a lady shooter might prefer a 28-gauge for its lightness and low recoil, the Trap shooter is bound to use a 12-gauge, and the wildfowler might demand a 10-gauge or even fit a swivel-mounted 4-gauge into his punt. But by and large 12-gauge is the universally accepted norm, with 20-gauge close behind. In the past there have been intermediate bores such as 8, 14, 16, 24 and 32-gauge, but these have all but disappeared and ammunition is no longer made for some of them. These alternatives were usually intended to produce a gun which had a heavier loading or was lighter or had some other notional advantage over the usual 12s, 20s and 28s, but advances in technology have now produced 12-gauge guns as light as the old 16-gauge and capable of firing as powerful a loading as most people can master.

It is, after all, the shot which kills the game, and the amount of engraving on the gun has no bearing on the matter. Nor, really, does the length of barrel, though some shooters are of the opinion that longer barrels develop better velocity; experiments show that most shot charges develop all their velocity in a 24-26 inch barrel and that anything more than that gives no advantage. Where the length of barrel counts is in the balance and feel of the gun, and the selection of barrel length (where a choice exists) should be based on that and not on any ballistic consideration.

What is perhaps more important than velocity is the pattern which the shot takes up during its flight. The shot is closely bunched as it leaves the muzzle but thereafter begins to spread out until at about 50 yards from the muzzle the dispersion is such that a bird could well survive as the shot passes all around it. At that range a kill is ballistically possible; the generally accepted energy level required in the shot pellet to kill a medium-sized bird is 0.85ft/lbs, and at 50 yards the pellets have more than this left in them. But the pattern density is the ruling factor. Adopting high velocity cartridges or increasing the weight of shot to try and compensate merely means more recoil and, in all probability, battering the gun to pieces without gaining any material advantage.

Pattern is governed by the propelling charge, the construction of the cartridge, and by the contours of the gun barrel. It is a well-known fact that the old-time black powder gun often shot a better pattern than today's smokeless powder gun, simply because of the different explosion characteristics of the propellants. Black powder 'pushed' the shot out far less violently than does smokeless, though, in theory at any rate, the reverse should be true. The wadding behind the shot, which transfers the powder's impulse, must be carefully arranged so that it does not force its way into the shot and disrupt the pattern, and the choke, or constriction in the barrel close to the muzzle, must be carefully contoured so that the shot leaves the barrel in the desired form. It is always worth trying different makes of cartridge in a gun to determine what difference there may be in patterning and to find out which combination gives the best result.

There is one other field in which the shotgun is gaining ground; that is the military and police employment. Shotguns are not generally considered to be military weapons but, when one thinks about it, they do have some definite advan-

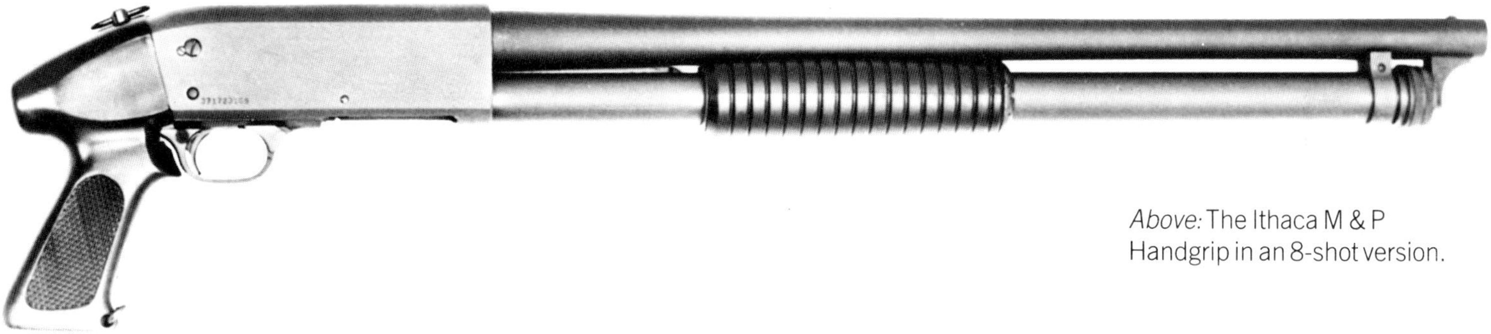

Above: The Ithaca M & P Handgrip in an 8-shot version.

Above: The radically modern 'bullpup' profile of the Mossberg Model 500 12-gauge.

tages in certain types of combat; as a quick-response weapon when surprised by an ambush, for example, the shotgun is an excellent device. They have been used, off and on, ever since the First World War, but in general their employment has been restricted because of the general unsuitability of commercial weapons. A side-by-side double gun is too long and too delicate for the rough-and-tumble of combat, and, of course, has only two shots in it, after which the soldier has to re-load. To be considered as a military weapon the shotgun must have some reserve of ammunition, must be short and handy, and must, above all, be robust.

The slide-action gun, with a short barrel, is the obvious choice and it is this type which has been extensively adopted by police and security forces and, to a lesser extent, by military users. The automatic shotgun is obviously a more attractive solution, if it can be made sufficiently rugged to stand service life, and this, too, has attracted a number of manufacturers. But fuller military acceptance depends largely upon the military users deciding what performance they require and then setting about designing a suitable weapon or, at least, financing some manufacturer to design one. No commercial company can afford a long and expensive program to produce a specifically military weapon without the prospect of a secure contract at the end of it, and for this reason the number of purely military designs which have appeared has been small.

One such government-funded design is the American 'Close Assault Weapon System' currently being developed. This demands a 12-gauge gun firing automatic and with an effective range of 100-150 yards, which obviously demands some fundamental research into ammunition as well as into the gun design. One result which has been disclosed is a weapon jointly developed by Olin/Winchester of America and Heckler & Koch of Germany. This fires a special belted 12-gauge brass-cased round which is loaded either with buckshot or with some 20 flechettes – small steel arrows – and

either loading appears to be lethal out to 150 yards range. Moreover the design of the extra-long case, with its raised belt in front of the rim, means that there can be no danger of someone loading this high-powered cartridge into a conventional shotgun and blowing it up.

Another American design, privately-funded, and which we cannot at this time name, uses an 8-shot rotary drum magazine which can be dropped out of the gun when all the shots have been fired and replaced by a fresh loaded magazine. The weapon does not eject the cases as it fires, so simplifying the mechanism, and changing magazines is a matter of seconds only. As designed it uses commercial 12-gauge cartridges, but the magazine could easily be chambered for CAWS ammunition or any other kind which might be developed.

Drum magazines are becoming popular; there is a South African design which uses a 12-shot drum, though this ejects the spent case after each shot and allows the drum to be re-loaded while on the gun. The only real drawback to the drum is its size; a 12-shot drum is a bulky thing to hang under a relatively short gun, but the box magazine, which is theoretically much better, is a difficult proposition for shotgun cartridges due to their squared-off front end which does not lend itself to smooth feeding from a box magazine.

Ammunition for military and police shotguns covers a much wider field of applications than does commercial shotgun ammunition. Not only are there the common-place buckshot loads, there are also flechette loads, tear gas pellets, rubber balls and bullets for riot control and illuminating, smoke and flare projectiles for signalling. One enterprising manufacturer even makes a message container which can have a written message tucked inside it and then be fired from a shotgun to land about 200 yards away. And how can we follow that?

Below: The classic lines of the Winchester Model 23 12-gauge.

SPECIFICATIONS OF REPRESENTATIVE MODERN SHOTGUNS

There are innumerable variations of some models; for example, the Browning B26 is available in two Skeet, two Game, two Game De Luxe, one Trap De Luxe and one Sporting version, and the differences between them do not lend themselves to simple tabulation. We have therefore confined the listings to one or two models to give an idea of the variety available. Note too that overall length is not a value which makers always tabulate.

NAME	TYPE	GAUGE	CHAMBER	LENGTH	WEIGHT	BARREL	SHOTS	REMARKS
ARMSEL: Armsel, Johannesburg 2000, Republic of South Africa								
Striker	Repeater	12	3in				12	Rotary magazine
A.Y.A.: Aguirre y Aranzabal, Eibar, Guipuzcoa, Spain								
Model 25	S/S	12	2¾in		6.3lbs	25.0in	2	Double trigger
Model 37 Super	O/U	12	2¾in		6.9lbs	26.0in	2	Also 16, 20, 28ga and .410
BERETTA: Pietro Beretta SpA, I-25063 Gardone Val Trompia, Brescia, Italy								
RS200	Slide	12	2¾in	40.5in	6.6lbs	20.5in	5+1	
SO2	O/U	12	2¾in		7.0lbs	28.0in	2	
AL2	Auto	12	2¾in		6.9lbs	26.0in	3	Also 20ga
S55B Trap	O/U	12	2¾in		7.2lbs	28.0in	2	Single trigger
C.B.C.: Companhia Brasileira de Cartouchoes, São Paolo, Brazil								
Model SB	S/S	12	2¾in	44.5in	6.6lbs	28in	2	Also 20ga and .410 (26in bbl)
FABRIQUE NATIONALE: Fabrique Nationale Herstal, B-4400 Herstal, Belgium								
Riot model	Slide	12	3in	38.2in	6.5lbs	19.7in	5+1	
Browning B125	O/U	12	2¾in		6.6lbs	27.5in	2	Single trigger
Browning FCS25	O/U	12	2¾in		6.6lbs	28.0in	2	Single trigger
Browning B27	O/U	12	2¾in		6.8lbs	27.0in	2	Single trigger
Browning B80	Auto	12	3in		7.7lbs	28.0in	4	Gas action. Also with 24, 26 and 30in barrels
Browning Auto 5	Auto	12	3in		7.7lbs	28.0in	5	Recoil action. Also with 26, 30 and 32in barrels
Browning BPS	Slide	12	3in		7.8lbs	28.0in	4	

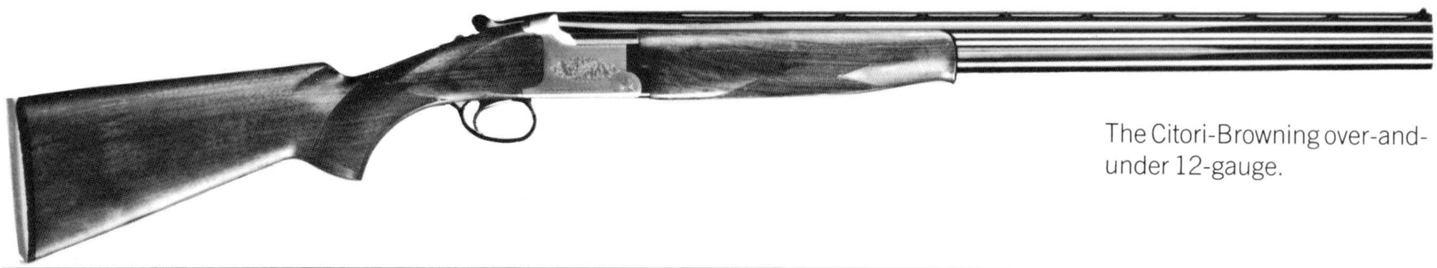

The Citori-Browning over-and-under 12-gauge.

NAME	TYPE	GAUGE	CHAMBER	LENGTH	WEIGHT	BARREL	SHOTS	REMARKS
FRANCHI: Luigi Franch SpA, I-25020 Fornaci, Brescia, Italy								
PA-3/215	Slide	12	2¾in	28.1in	6.0lbs	8.7in	3	Twin handgrips, no stock, police and military only
PA3/345	Slide	12	2¾in	33.0in	6.2lbs	13.5in	5	Police
PA3/470	Slide	12	2¾in	38.2in	6.5lbs	28.5in	7	PA3/345 and 470 have folding butt-stock
PA-8	Slide	12	2¾in	32.0in	6.5lbs	24in	7+1	Combat style
SPAS-11	Auto	12	3in	35.4in	7.1lbs	19.7in	7+1	Auto or slide action at will
SPAS-12	Auto	12	3in	36.9in	9.25lbs	18.1in	7+1	as 11 but with folding stock
Model 500	Auto	12	2¾in		7.2lbs	26.0in	5	Gas action
Hunter	Auto	12	2¾in		7.5lbs	26.0in	5	Recoil action
Falconet	O/U	12	2¾in		6.7lbs	26.0in	2	
HARRINGTON & RICHARDSON: Harrington & Richardson Inc., Gardner, MA 01440								
Model 088	Single	12	3in	46.0in	6.0lbs	30.0in	1	Also 16, 20, 28ga and .410
Model 099 De Luxe	Single	12	3in	43.0in	6.0lbs	28.0in	1	Also 16, 20ga and .410
Model 176 Slug	Single	10	3½in	44.3in	9.2lbs	28.0in	1	
Model 490	Single	20	3in	40.0in	5.0lbs	26.0in	1	Also .410
HECKLER & KOCH: Heckler & Koch GmbH, Oberndorf-a-Neckar, West Germany								
CAWS	Auto	12	Special	30.0in	9.0lbs	18in	10	Development

The Heckler & Koch 'Close Assault Weapon' automatic shotgun.

NAME	TYPE	GAUGE	CHAMBER	LENGTH	WEIGHT	BARREL	SHOTS	REMARKS
ITHACA: Ithaca Gun Co., Ithaca, NY 14850								
Model 37 M&P	Slide	12	2¾in		6.5lbs	18.5in	5	Also 7-shot with 20in bbl

Above: The basic 5-shot version of the Ithaca Model 37 M & P.

Above: The Model 37 Deerslayer. Note the modified sights compared with the basic version at the top of the page.

Above: The Model 37 with pistol grip and stock.

NAME	TYPE	GAUGE	CHAMBER	LENGTH	WEIGHT	BARREL	SHOTS	REMARKS
Handgrip	Slide	12	2¾in		5.2lbs	18.5in	5	Stockless version of 37
LAPD	Slide	12	2¾in		6.5lbs	18.5in	5	
Stakeout	Slide	20	2¾in		3.5lbs	13.2in	5	Stockless light weight; also 12ga version
Mag-10 Roadblocker	Auto	10	3½in		10.7lbs	22.0in	3+1	Gas operated
LANBER: Armas Lanber SA, Zaldibar, Spain								
Model 844	O/U	12	2¾in	44.4in	7.7lbs	27.7in	2	Single trigger
MANUFRANCE: Manufrance, St Etienne, France								
Auto	Auto	12	3in		6.8lbs	28.0in	3	Also 26, 30in barrels. Gas action
MAROCCHI PIERETTI: Stefano Marocchi i Figli, I-25063 Gardone Val Trompia, Brescia, Italy								
Non-Ejector	O/U	12	2¾in		7.0lbs	27½in	2	Double trigger
Ejector	O/U	12	2¾in		7.0lbs	28in	2	Single trigger
Contrast Trap	O/U	12	2¾in			29½in	2	Single trigger

NAME	TYPE	GAUGE	CHAMBER	LENGTH	WEIGHT	BARREL	SHOTS	REMARKS
MIROKU: Miroku Firearms Mfg. Co., Kochi-Shi, Kochi-Ken, Japan								
Model 3000	O/U	12	2¾in		6.9lbs	28.0in	2	Single trigger
Model 4800	O/U	12	2¾in		7.5lbs	29.5in	2	Single trigger
Model 6000 Skeet	O/U	12	2¾in		7.5lbs	27.0in	2	Single trigger
Model 7000	O/U	12	2¾in		7.9lbs	29.5in	2	Single trigger
MOSSBERG: O. F. Mossberg & Sons Inc., North Haven, CT 06473								
Model 595	Single	12	3in	45.7in	7.5lbs	28.0in	3	Bolt action
Model 500ATP6	Slide	12	3in	37.7in	6.5lbs	18.5in	6	
Model 500ATP6C	Slide	12	3in	28.0in	5.7lbs	18.5in	6	Stockless version of ATP6
Model 500 ATP8	Slide	12	3in	39.7in	6.7lbs	20.0in	8	Also ATP8C stockless version
Model 500 Bullpup	Slide	12	3in	30.8in	8.5lbs	20.0in	8	
Model 5500	Auto	12	2¾in	38.5in	7.5lbs	18.5in	5	Also stockless version
NORINCO: China North Industries Corp., Beijing, China								
HL-12-101	Single	12	2¾in	45.2in	5.7lbs	29.5in	1	
HJ-12-201	O/U	12	2¾in	46.7in	8.8lbs	29.5in	2	Also 27in bbl
HL-12-202	O/U	12	2¾in	46.2in	7.9lbs	29.5in	2	Single trigger
PERUGINI: Fabbrica Italiana Armi Pietro Perugini, Nuvolera, Brescia, Italy								
Emperor	S/S	20	2¾in	44.7in	6.6lbs	27.8in	2	
REMINGTON: Remington Arms Co., Bridgeport CT 06601								
Model 870 Spl Field	Slide	12	3in	41.0in	6.8lbs	21in	3+1	Also 20ga
Model 1100	Auto	12	2¾in		7.5lbs	28.0in	3	

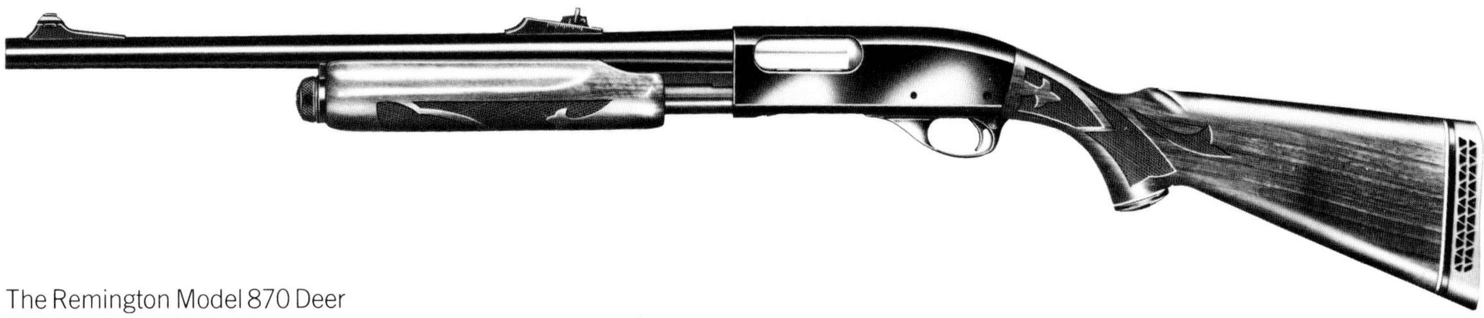

The Remington Model 870 Deer Gun with left hand action.

NAME	TYPE	GAUGE	CHAMBER	LENGTH	WEIGHT	BARREL	SHOTS	REMARKS
RENATO GAMBA: Armi Renato Gamba SpA, Gardone Val Trompia, Brescia, Italy								
Hunter	S/S	12	2¾in		6.8lbs	27.5in	2	
Oxford Competition	S/S	12	2¾in		6.8lbs	27.5in	2	
London Royal	S/S	12	2¾in		6.3lbs	26.7in	2	
Ambassador	S/S	20	2¾in		6.4lbs	27.5in	2	
Dedalo	O/U	12	2¾in		7.0lbs	28.0in	2	
Edinburgh	O/U	12	2¾in		7.0lbs	28.0in	2	Single trigger
Victory Trap	O/U	12	2¾in		7.7lbs	29.1in	2	Single trigger
Super Model 80	Auto	12	2¾in		7.0lbs	28.0in	3-7	Gas
ROSSI: Amadeo Rossi y Cia, 93000 São Leopoldo, RGS, BraZIL								
Squire	S/S	12	2¾in		7.8lbs	28.0in	2	Also 20ga and .410

The Rossi Squire, a traditional 12-gauge design.

NAME	TYPE	GAUGE	CHAMBER	LENGTH	WEIGHT	BARREL	SHOTS	REMARKS
RUGER: Sturm Ruger & Co., Southport CT 06490								
Red Label	O/U	12	2¾in	43.0in	7.5lbs	26.0in	2	Also 20ga

The Ruger 12-gauge.

NAME	TYPE	GAUGE	CHAMBER	LENGTH	WEIGHT	BARREL	SHOTS	REMARKS
SIDEWINDER: Sage International Ltd., Rochester MI 48063								
1100 Sidewinder	Slide	12	2¾in		8.0lbs	18.8in	7	Modified Remington 1100 with folding butt, etc.
S.K.B.: S.K.B. Arms Co., Chuo-ku, Tokyo, Japan								
M-7300	Slide	12	2¾in		7.5lbs	28.0in	4	Also 20ga
XL-900	Auto	12	2¾in		7.8lbs	28.0in	4	Gas action

NAME	TYPE	GAUGE	CHAMBER	LENGTH	WEIGHT	BARREL	SHOTS	REMARKS
SMITH & WESSON: Smith & Wesson, Springfield, MA 01101								
Model 916	Slide	12	3in		7.3lbs	28.0in	6	
Model 1000	Auto	12	3in	48.0in	7.5lbs	28.0in	4	
Model 3000 Police	Slide	12	2¾in	38.5in	6.8lbs	18in	5 or 8	

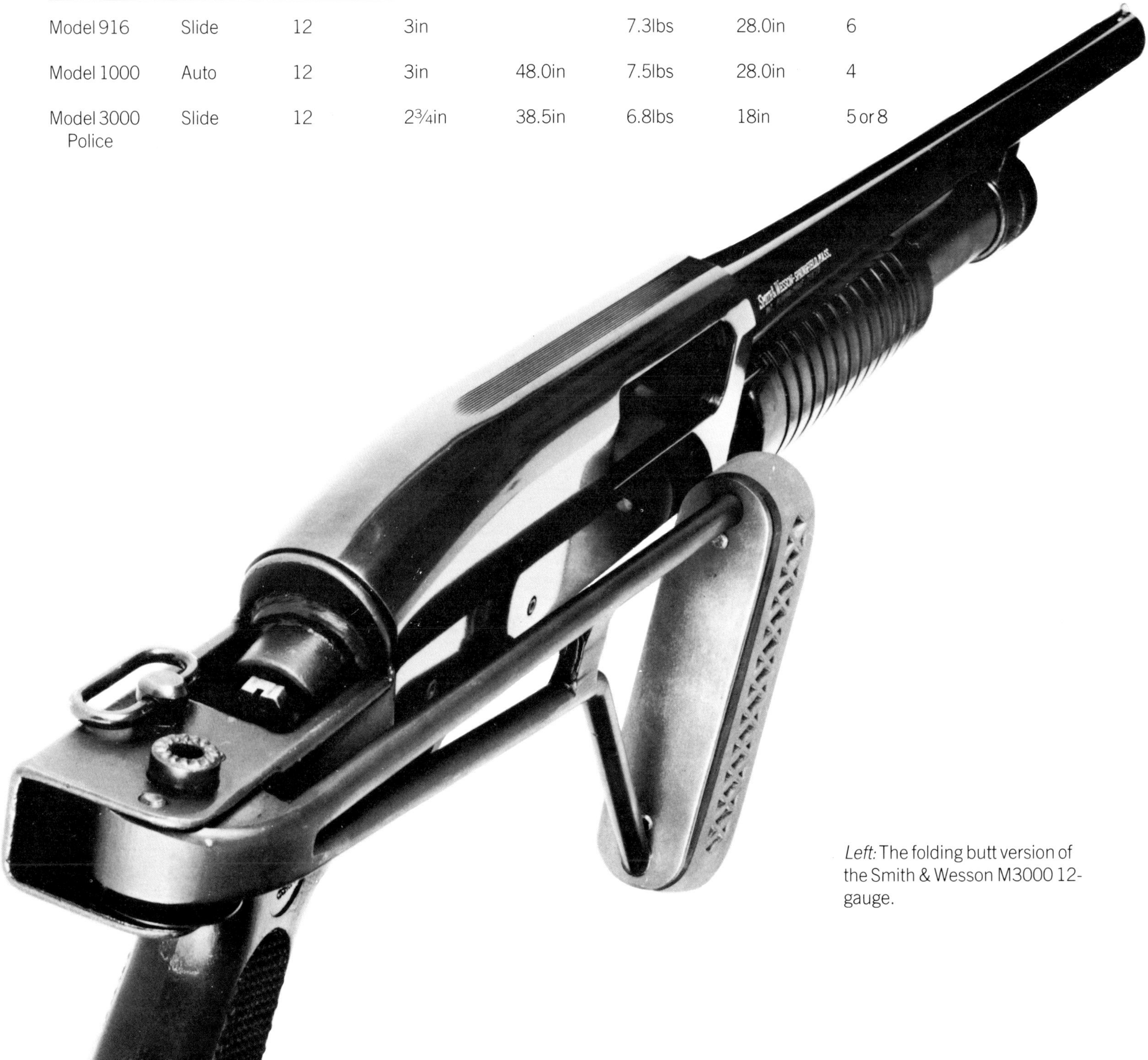

Left: The folding butt version of the Smith & Wesson M3000 12-gauge.

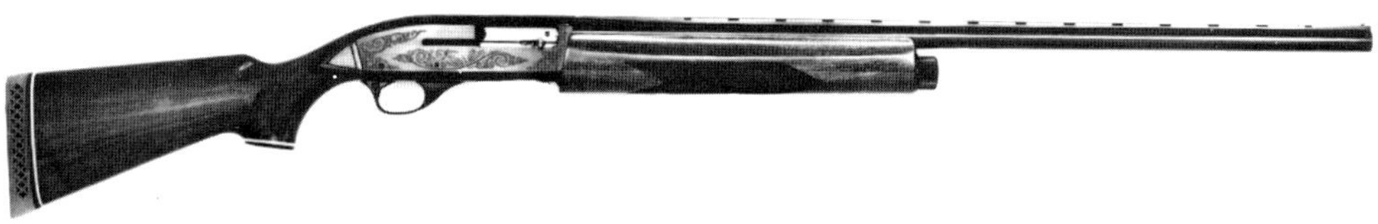

Above: The Smith & Wesson M1000, chambered for the 3-inch Magnum cartridge.

UGARTACHEA: Ignacio Ugartachea, Eibar, Guipuzcoa, Spain

NAME	TYPE	GAUGE	CHAMBER	LENGTH	WEIGHT	BARREL	SHOTS	REMARKS
Model 30	S/S	12	2¾in	45.5in	6.9lbs	28in	2	Double trigger

VIKING: Viking Arms Ltd., Harrogate, England

NAME	TYPE	GAUGE	CHAMBER	LENGTH	WEIGHT	BARREL	SHOTS	REMARKS
SOS	Slide	12	2¾in		7.4lbs	24¼in	7+1	Combat style

Below: The Viking Police Model 12-bore with slide action.

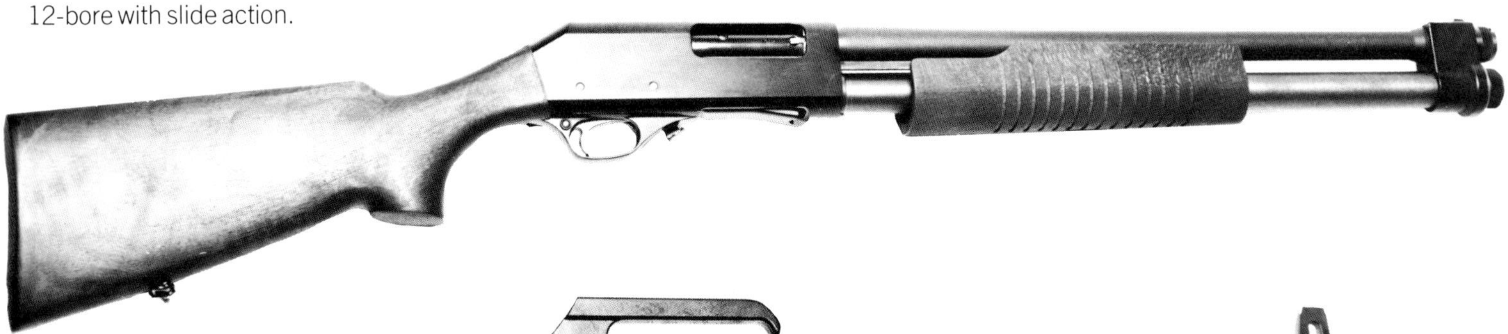

Above: A Viking semi-automatic 12-bore, designed for military use.

WEATHERBY: Weatherby Inc, South Gate, CA 90280

NAME	TYPE	GAUGE	CHAMBER	LENGTH	WEIGHT	BARREL	SHOTS	REMARKS
Model 82	Auto	12	2¾in	48.5in	7.7lbs	28.0in	5	Gas action
Model 92	Slide	12	2¾in	48.5in	7.5lbs	28.0in	5	

WINCHESTER: US Repeating Arms Co., Postbox 30-300, New Haven, CT 06511

NAME	TYPE	GAUGE	CHAMBER	LENGTH	WEIGHT	BARREL	SHOTS	REMARKS
Defender	Slide	12	3in		6.7lbs	18.0in	6+1	
Stainless Police	Slide	12	3in		7.0lbs	18.0in	5+1	
Stainless Marine	Slide	12	3in		7.0lbs	18.0in	6+1	
Model 21	S/S	12	Dimensions depend upon customer specification					
Model 23UK	S/S	12	2¾in		6.3lbs	28in	2	
Model 1300	Slide	12	3in	42.6in	6.4lbs	22.0in	5	Featherweight

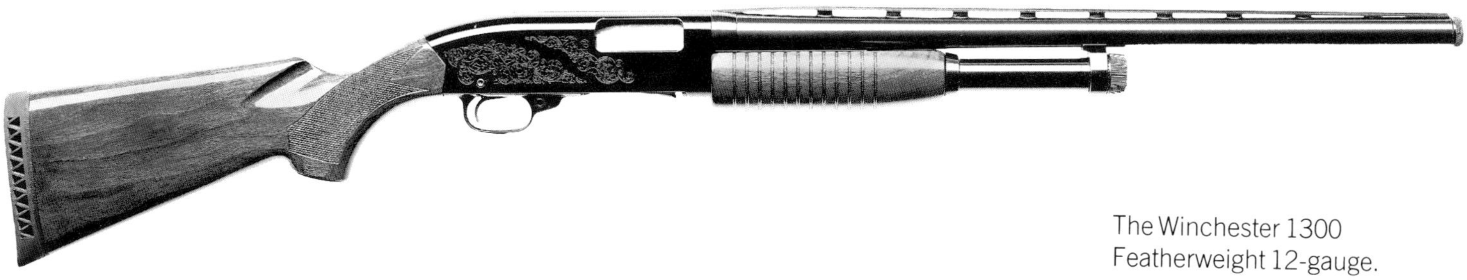

The Winchester 1300 Featherweight 12-gauge.

The Winchester Ranger semi-automatic shotgun.

WINCHESTER Contd.

NAME	TYPE	GAUGE	CHAMBER	LENGTH	WEIGHT	BARREL	SHOTS	REMARKS
Ranger	Slide	12	3in	48.7in	7.3lbs	28.0in	5	Also 20ga
Ranger	Auto	12	2¾in	48.7in	7.3lbs	28.0in	3	Gas. Also 20ga

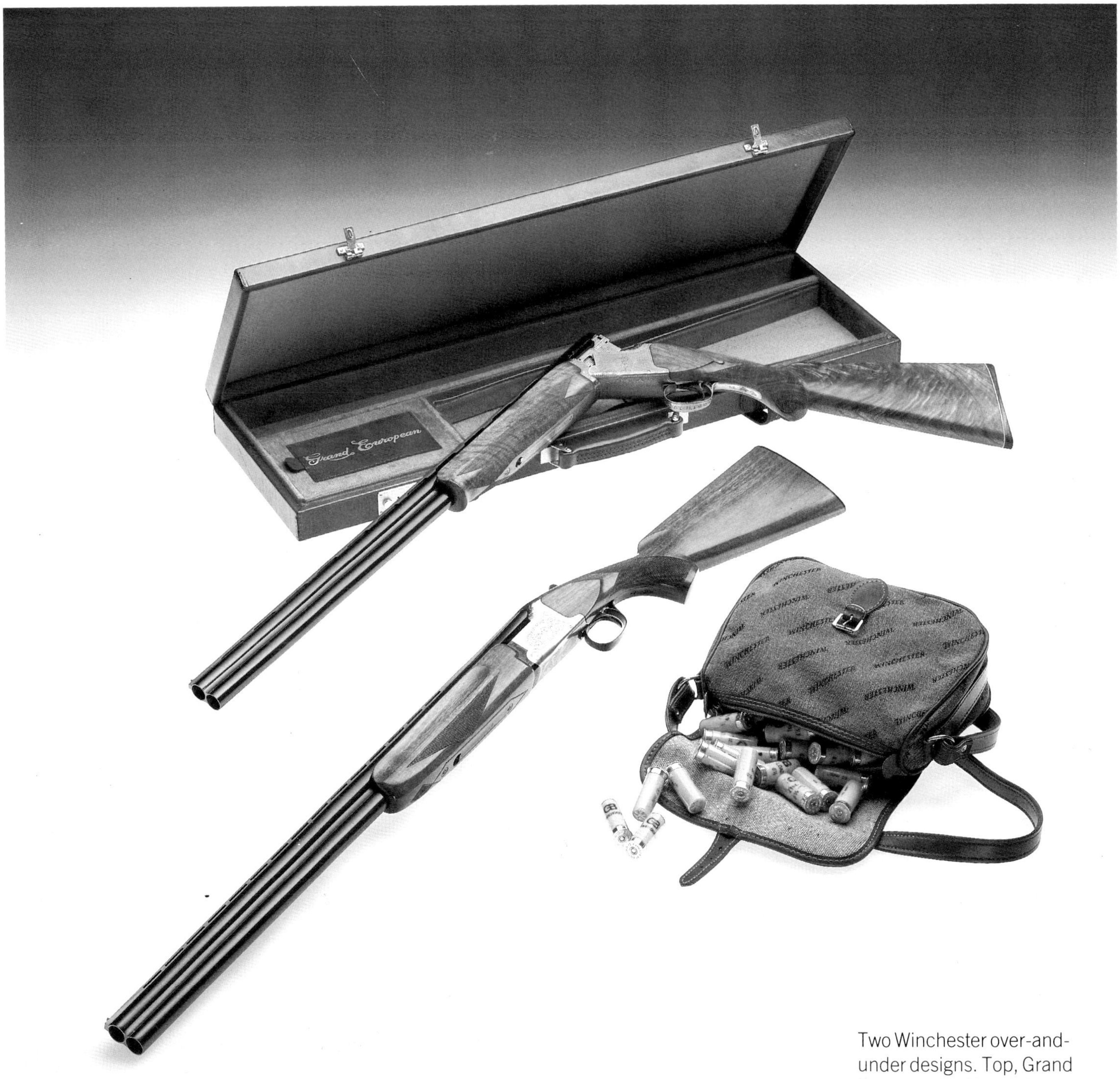

Two Winchester over-and-under designs. Top, Grand European Field and, lower, Lightweight Field.

The well known Browning High Power automatic in 9mm Parabellum. This is the Mark 2 version with ambidextrous safety catch.

PISTOLS

There are three types of pistol in current use; the revolver, the semi-automatic, and the 'free pistol', a catch-all classification which includes a variety of weapons developed solely for competitive shooting in certain highly-developed types of contest.

With one or two exceptions, the *revolver* has changed little during the course of the present century. Two types predominate; the hinged frame model in which the barrel and cylinder hinge down to expose the chambers for re-loading, and the solid frame type in which the cylinder is swung out sideways, on a 'crane', to reload. Both have their advocates and both systems have adequate strength for the heaviest loads when built sufficiently robustly, but today the tendency is to use the solid frame system for the more powerful weapons, the hinged frame being generally found on small-caliber sporting revolvers. This appears to be a function of weight and manufacturing simplicity; all things being equal, a powerful caliber revolver on the hinged frame principle will be heavier than that on the solid frame, due to the need to ensure adequate strength in the locking of the frame, and the manufacture of a one-piece frame by modern methods is cheaper and faster than that of a two-piece construction.

There are, though, other methods of construction; the gas-seal revolver still finds favor in Eastern Europe, for example. This system seeks to remedy the principal theoretical disadvantage of revolvers, the fact that there must be a physical gap between the front face of the chamber and the rear face of the barrel (otherwise the cylinder would not revolve freely). This, in turn, means that there will be a leak of propellant gas from this joint at the time of firing, and the gas-seal revolver seeks to seal this gap by mechanical means so as to prevent the leak of gas and so ensure that the entire power of the propelling charge is usefully expended in driving the bullet up the barrel.

This sealing is done by a combination of mechanical design in the gun and in the cartridge. The cartridge case is longer than normal and extends to the tip of the completely-enclosed bullet. The cylinder is free to move longitudinally and is controlled by a pressure plate forming part of the trigger-hammer linkage. Finally the rear end of the barrel bore is slightly enlarged so that the mouth of the cartridge case can enter, while the external surface of the barrel end is coned, and the mouth of the chambers belled, so that the two can mate. As the hammer of the gas-seal revolver is cocked, so the cylinder is forced forward until the cartridge case mouth enters the rear of the barrel bore, while the mouth of the chamber slides tightly over the coned end of the barrel exterior. The cylinder is now locked in this position, with a

Below: A highly decorated Renato Gamba Trident .38 Special.

Above: The unusual and radical profile of the Mateba design.

recoil plate thrust firmly behind it to prevent movement of either the cylinder or the cartridge case, and the hammer falls and fires the round. The bullet passes from the cartridge case into the bore, the case metal and the mating surfaces of the cylinder and barrel forming a more-or-less perfect seal to prevent any leakage.

The gas-seal revolver is not new; it was developed in the 1880s and found its resting place in the Russian 7.62mm Nagant service revolver of the 1890s, a revolver which remained in Tsarist and Soviet service until after World War II. But attractive though the theory is, it was never embraced anywhere else, and only the Russians and their satellites have retained it for competition revolvers today. The obvious question to be asked is 'Is all this mechanical complication worth the effort?'

The answer has to be somewhat guarded; it seems to be 'Yes, in a competition weapon'. Most examinations into the efficacy of gas seal revolvers has been devoted to studying the relative improvement in muzzle velocity, since this is the most obvious area of likely improvement. Unfortunately, it is impossible to find two revolvers, one gas-seal and the other not, which are otherwise absolutely identical and could thus be fired in comparative trials. Moreover, experience suggests that even two identical revolvers, be they gas seal or not, will rarely shoot the same, since there are several variables which enter into the equation. But it seems to be generally accepted that the gas sealing system on the Russian service revolver probably made a difference of some 70 feet per second in its muzzle velocity. For combat purposes, this is neither here nor there; it probably improves the striking energy by some 35 foot-pounds, but since this means a muzzle energy of about 40 foot-pounds, more than sufficient to cause injury, the difference is academic.

In a competition revolver, however, there may be some advantage; here muzzle velocity is less important than consistency. The object of competitive shooting, after all, is to put the shots into as small an area of the target as possible, and once the sights of the weapon have been adjusted to suit the velocity of the chosen bullet the principal demand is that every bullet thereafter should have the same velocity so as to follow the same trajectory and strike in the same spot. Conventional revolvers have gaps between the cylinder and barrel which vary widely from gun to gun; from as small as 0.0015 inch to as large as 0.006 inch. Tests carried out several years ago showed that at the moment of maximum chamber pressure there is no leak, since the bullet is in process of crossing the gap and sealing it, but that immediately after this the chamber pressure drops very rapidly. This is because the bullet is now being forced into the rifling, and therefore the sudden check to its progress allows leakage to occur. Minute differences in bullet diameter, bullet length, seating depth in the cartridge case, all combine to make the amount of gas leakage variable from shot to shot, and therefore the eventual muzzle velocity will never be constant, swinging as much as 50 feet per second from shot to shot. With this sort of variation, consistency of strike is impossible. The gas seal revolver, however, holds its velocity to a much closer spread, no more than 10-15ft/sec (since the other variables still apply) and must therefore produce tighter groups at the target.

A totally different approach to revolver design and construction appeared in 1983 with the Italian 'Mateba' revolver, the invention of Sr. Amelio Ghisoni of Pavia. The Mateba has moved the cylinder from its traditional position above the trigger to a location in front of it, the cylinder axis being level with the trigger finger. In broad terms the pistol is of solid frame construction, with a swing-out cylinder, but the area behind the cylinder more closely resembles an automatic pistol, having a rectangular frame and a more-or-less straight butt instead of the rounded butt of a conventional revolver. The result of this redesign is that the axis of the barrel lies much closer to the firer's hand, the prolongation of the bore passing no more than about one inch above the top of the hand as it grips the butt. A second result is a displacement of the center of gravity to somewhere about the center of the cylinder when the gun is loaded. This means that the balance characteristic is unlike any other handgun; revolvers are usually top-heavy and automatics butt-heavy; the Mateba tends toward muzzle-heaviness but in fact feels evenly-balanced when in the hand, leading to a very steady shooting hold. The lowset design means that there is less tendency to allow the weapon to roll sideways, as can happen with revolvers, and there is much less tendency for the muzzle to rise during firing, due to the more favorable leverage couple between the axis of the bore and the point of hold. The recoil does not lift the gun and cause the frame to bite into the web between thumb and forefinger, but seems to drive the weapon back more squarely and thus spread the recoil across the palm of the hand.

The firing mechanism is selective single or double action, using an internal hammer working against a long firing pin supported in the frame. There is an external cocking lever on each side of the frame to permit single-action cocking or to allow controlled release of the hammer, and the trigger pull can be adjusted by an external screw. The cylinder, positioned as it is, can be larger than is normal, and the standard Mateba in .38 caliber uses an 8-shot cylinder. The unusual feature is that the cartridges are pre-loaded into a form of 'speed loader', a perforated metal plate which holds each cartridge just ahead of the rim; a second plate, perforated to permit the passage of the firing pin, then hinges over and locks the cartridges in place. The entire unit is simply loaded into the cylinder as one item, and the loading plate remains there as part of the assembly; it is not necessary to detach the cartridges from the loader.

An interesting development from the basic revolver design is the MTRC carbine; this uses the grip, frame and cylinder assembly of the revolver but allies it to a longer barrel and a pressed-steel buttstock to produce a shoulder arm. Two models have been developed, the MTRC 8 with a 300mm barrel and the MTRC 8L with a 460mm barrel. There is a plastic fore-end to provide grip and also an ingenious sheet-metal guard which hinges up to wrap around the cylinder and thus protect the supporting arm from any gas leaks from the chamber/barrel junction.

Automatic pistols appear to be undergoing a short rest period after a decade of furious activity on the design front. The past ten or twelve years has seen a profusion of new designs, particularly in the military field, and it now seems that there will be a period of assimilation, while these new models are brought into use and analysed by user experience. The broad outlines of design have largely been similar: large magazines, double-action lockwork, ambidextrous safety catches and magazine releases have been adopted almost universally, largely as a result of two major competitions, the US Army's search for a new service pistol and the West German police request for a foolproof police pistol. At the time of writing (July 1984) the solution to the American problem has not yet been announced, though in our view the ultimate winner will be the SIG-Maremont P226 pistol. The German contest took place several years ago and the result of that can be seen in the Heckler & Koch P7 series. The reasons for these selections, though, are by no means similar and the demands are worth some study.

The German police requirement was for a pistol in 9mm, .38 or .357 caliber, with a minimum muzzle energy of 369 foot-pounds, a minimum muzzle velocity of 1150 feet per second, and a minimum magazine capacity of 6 shots. It had to weigh no more than 1 kilogram (2lbs $3\frac{1}{4}$oz), be no more than 7 inches long, $5\frac{1}{8}$ inches high and $1\frac{3}{8}$ inches wide and with a barrel length of 8 to 10 calibers. The lock had to be double action for the first shot, single action thereafter, and that first shot had to be fired without the need to operate any form of safety catch. The gun had to be safe against any form of accidental discharge, had to fire 5000 rounds without failure of any component, and had to put all its shots into a 6 inch circle at 25 yards range. There were several other specified points covering such items as grip angle, cylinder/barrel gap, ease of maintenance, sights and so forth, but the foregoing features were the most significant ones.

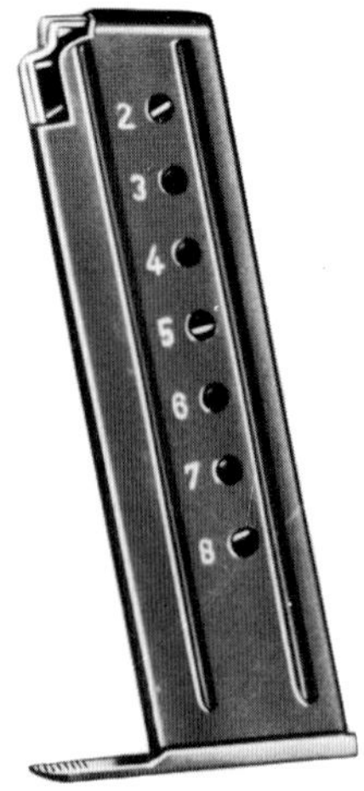

The Heckler & Koch P7 reduced to its basic components. Note the spring guide rod attached to the slide which delays breech opening by acting as a recoil buffer.

This page: Three views of the Austrian Glock 17.

On close examination, this apparently nonpartial specification indicated that nobody was taking the idea of a revolver very seriously; it was unlikely that any revolver could deliver the desired 369 foot-pounds ME with a cylinder diameter of only 1⅜ inches, the maximum specified width. Moreover the specification also stated that whatever pistol was adopted, it would have to fire common ammunition with existing submachine guns. Moreover, by a strange coincidence, the energy and velocity values were precisely those of the service 9mm Parabellum cartridge, so it didn't require a Sherlock Holmes to realise the German police had set their hearts on that caliber.

Four manufacturers faced up to this requirement; Heckler & Koch, Walther, JP Sauer & Sohn and Mauserwerk. H&K put forward their existing P9S, admitting it was oversized but promising a new and improved design to follow; Walther offered the P4, a shortened version of the familiar P38 with an added decocking lever; Sauer came up with their SIG-Sauer P220, again over-sized but with the promise of a smaller model to come; and Mauser developed something totally new called the HsP. These were all submitted to firing trials and environmental testing; all failed. Mauser withdrew from the contest, Sauer produced the P225, Walther produced the P5, a redesign based on the P38 but aimed at filling all the physical requirements, and Heckler & Koch produced their radical P7 (or PSP as it was then known). More trials followed and the P225 was the only one to survive.

For some unknown reason a further trial was convened, at which the Walther P5 passed successfully, and then a fourth series of trials got the P7 into the success group. It now meant that three trials had managed to get all three contestants past the winning post, and in a masterpiece of open-mindedness the test board, in their report, said that since all three had passed, all were therefore approved for police use and it was thereafter up to individual police forces to select which of the three they preferred. The P225 was formally rechristened the Pistole 6, the PSP became the Pistole 7 (the Walther was already the Pistole 5), and that was that.

The American test procedure was totally opposite; instead of endeavoring to get everybody past the post, the object appears to have been to wipe every contestant out on some excuse or other. The full gamut of conventional tests was gone through - accuracy, reliabity, mud, dust, water, freezing, baking – and anything which survived had its grips removed and was clamped into a vibrator unit and vibrated at a high frequency until components fell out; it was then eliminated from the trial. The official justification for this was simply that it was possible to do this with a Colt M1911

Heckler & Koch P9S. Note the decocking lever behind the trigger guard.

The Mauser revolver in .38 Special caliber.

without it suffering, therefore anything destined to replace the M1911 had to be able to survive it as well. Logical as this may have seemed to the man who thought it up, it neglected the fact that the US demands for such things as firing pin safeties, ambidextrous magazine catches and de-cocking levers meant that the frame was full of pieces which weren't present or even thought of in the M1911 design.

What tended to upset people about the US trials was simply that most of the contestants were of foreign origin; Beretta (which did particularly well), SIG-Sauer, Heckler & Koch, Star and FN designs competed with Colt and Smith & Wesson. This merely stressed the imbalance of design effort; more European companies had been working on modern auto pistols than had US companies. What was more surprising to many onlookers was the number of pistols which were not tested, though this was probably due to the designers' reluctance to enter them into a trial they felt they had no hope of winning. The requirement, as well as laying down basic ballistic demands, also demanded that the selected pistol be manufactured in the USA, and therefore only those companies who either had a US affiliate or were within sight of organising one were willing to submit samples. Nevertheless, one might have expected the US authorities to put some of these other pistols in, merely to find out how they stacked up. The Austrian Steyr GB and Glock 17, the Italian Bernadelli, Tanfoglio and Benelli models, the Astra A80, the Llama Omni, any or all of these would have added some spice to the proceedings, though in fairness we would have to admit that it would also have dragged the affair out to an even greater length.

Another remarkable feature of the tests was what went on behind the scenes; for example, a report to the author *from behind the Iron Curtain* informed him that Aberdeen Proving Ground were testing a design from Sturm Ruger. Subsequent investigation appears to confirm this, but there has never been any public admission of this design or its testing, and one is inclined to wonder what others were given quiet tests at the same time.

In any event, it looks as if an announcement will be made before this book reaches print and, as we said earlier, our money is on the SIG-Maremont P226 for reasons which we would prefer not to state.

The SIG-Maremont P226 is, in fact, a slight re-hash of the SIG-Sauer P225 which, as we saw above, was a redesign of the P220. The P220 had to be changed in order to get within the size limitations of the West German Police specification; it then had to be changed again to meet the American demand for an ambidextrous magazine release and also the increased capacity of the magazine from 8 rounds of the P225 to the 15 rounds demanded by the US specification. In order to accede to the US demand for local manufacture, SIG and JP Sauer appointed the Maremont Corporation as their US factors,

The Smith & Wesson M549, one of the contestants in the US tests.

hence the 'SIG-Maremont' designation; Maremont have a good reputation for manufacturing the M60 and M2HB machine guns, so they are no strangers to the sort of precision demanded in the manufacture of military arms.

Irrespective of who wins the US contest, the most interesting fact is simply that at long last the grip of the .45 cartridge on American thinking has been loosened. Europeans have been trying to get the Americans to look at the 9mm Parabellum for forty years without much success; only the .45 was capable of stopping an adversary was the claim, which was hard to refute, even though 9mm Parabellum bullets have probably killed far more people than have .45 bullets. There are also justifiable objections to the general level of accuracy obtainable with the NATO standard 9mm Parabellum bullet, and it is understood that the US authorities are currently looking closely at this area in the hope of

The Czech CZ75 9mm automatic is highly regarded by experts in the West.

developing a new 9mm round which will improve accuracy and stopping power and be acceptable as a new NATO standard. After all, the current NATO bullet is the British Mark 1Z of 1941, which in turn was more or less copied from the German service bullet adopted in 1915, so one might be excused for suspecting that there is room for improvement.

It is not without interest, in this context, to look at the new 'Bren Ten' pistol developed by Jeff Cooper. Mr Cooper has long been one of the most outspoken critics of the 9mm Parabellum cartridge, but he was sufficiently open-minded to be able to admit that the .45 auto round was deficient in many respects. He therefore set about developing a 10mm (.40) caliber cartridge and a pistol to match, a combination which eventually appeared in 1983 as the 'Bren Ten'.

The 'Bren' part of the name comes from the adoption of the Czech CZ75 pistol's double action lockwork; this was developed at Brno, in Czechoslovakia, which was also the factory which originated the ZB26 light machine gun which later became better known as the British 'Bren Gun'. This is the official statement; it would ring better if the CZ75 was actually made in Brno, but in fact it is made in Uhersky Brod, which rather takes the edge off the story. Never mind, it makes for a nice euphonious name.

The pistol is a fairly conventional design of hammer-fired semi-automatic using the Browning type of short recoil

The Detonics Scoremaster in .45 caliber.

breech lock. With a 5 inch barrel, it weighs 39 ounces empty and has a magazine capacity of 11 rounds. The barrel is rifled to a patented design known as 'Power-Seal' in which the rifling grooves merge into the lands at a gentle angle rather than the conventional sharp 90° angle; in many respects it resembles the 'polygonal' rifling used by Heckler & Koch of Germany but employs five grooves rather than four. The advantage of this type of rifling is that less pressure is required to engrave the bullet into the grooves, the bullet is centered better in the bore, and the gentle angles lead to a better gas seal between bullet and bore. All this adds up to a degree of accuracy and consistency which is remarkable; 10-shot groups within 0.72 inch at 100 feet have been reported during testing of ammunition.

The cartridge has been developed by the Norma company and uses a straight-walled rimless case with a 200-grain full-jacketed bullet. Muzzle velocity from the 5 inch barrel is 1200ft/sec, giving a muzzle energy of 640 foot-pounds. Translating this into stopping power, using the scale developed by Colonel Hatcher many years ago, the 10mm Bren Ten gives a value of 72, while the 9mm Parabellum gives 28 and the .45 automatic gives 62. It is only exceeded by such powerful loadings as the .41 and .44 Magnum revolver rounds, so that the Bren Ten qualifies as the most potent conventional automatic pistol in production.

However, ballistics alone do not decide the question of adoption of a weapon, and the day has long gone when an inventor could produce a weapon of an odd caliber and then sit back to enjoy the royalties on the ammunition. To have any chance of serious adoption a pistol must use a cartridge of standard form, readily obtainable anywhere in the world; it will be recalled that the German pistol specification demanded that the weapon should use the existing submachine gun cartridge, and similar logistic considerations must apply to most organised forces. While we would not argue with the Bren Ten as a good ballistic solution, we have to say that it is extremly unlikely to see any widespread service adoption, simply on the grounds of the selected caliber. Which is grossly unfair to Jeff Cooper, but that's the way the world works.

In spite of the attractions of other calibers the .45 pistol still has many adherents and there are a number of companies who have simply set out to refine the basic Colt M1911 concept. The Colt was developed purely as a military weapon, and as such it was entirely satisfactory; but those who want to shoot competitively demand a somewhat better degree of accuracy, and those who enjoy guns for their own sake demand a rather more refined standard of finish. For this latter group, of course, Colt manufacture the pistol commercially as their 'Government Model' and offer it with a superlative finish. But even so, there are those who like something a bit different, and it is here that the specialist gunmakers find their market.

One such 'refined 1911' is the Arminex Trifire which, in the words of its designer James Mongello is 'the 1911 brought up to modern standards'. Whilst the pistol retains the general appearance and operating system of the M1911, it has been changed in several small but significant respects: the grip safety has been removed and the frame contour altered so that the spur now extends further over the firer's hand and there is no longer a danger of having the pistol cut the web between the thumb and forefinger; the frame-mounted safety has been replaced by a slide-mounted ambidextrous safety which securely locks the firing pin; the recoil spring system has been redesigned to simplify removal and also to offer smoother cycling; the slide stop lever has been extended so that it falls closer to the firer's thumb; and the hammer is changed from the long spurred type to a neater ring-spur. All small things, to be sure, but they make a distinctive change in the handling of the weapon. The trigger guard has been re-shaped to permit a more firm two-handed grip, the barrel is of stainless steel, and the sights are fully adjustable. The result of all this is a weapon which, while still of the familiar pattern, delivers extremly reliable and accurate performance allied to an excellent standard of fit and finish. The Trifire is also available in 9mm Parabellum and .38 Super Auto calibers for those who prefer them, and there are also Target Models with a 6-inch barrel and longer slide.

Another 'improved 1911' is the Detonics 'Scoremaster'. The Detonics company are well-known for their pocket combat pistols – cut-down versions of the basic 1911 pattern – but the 'Scoremaster' is a full-sized pistol intended for IPSC competition. Made of stainless steel throughout, it combines a patented self-centering cone barrel system and a compound recoil spring assembly which, together, ensure smooth recoil characteristics and a consistency in barrel centering which ensures that the barrel and sights are always correctly related to each other. With the standard 1911 pattern barrel assembly the barrel bush (in front of the slide) is not a precision fit and is subject to wear, so that the relationship of barrel to slide and sights is not consistent. Detonics' cone assembly replaces the barrel bush with a unit which always brings the barrel back to the same place, compensating for gradual wear. Other features on the Scoremaster include a National Match pattern trigger, ambidextrous safety, fully adjustable rear sight and wrap-around combat style grip.

Detonics have also developed a totally new cartridge, for which the Scoremaster can be chambered instead of the standard .45 ACP. This is the .451 Detonics Magnum cartridge. Slightly longer than the .45 ACP, the .451 is based on the case of the 7.62×51mm (.308 Winchester) rifle cartridge, trimmed and necked to take a .451 bullet. Various bullets and loadings have been developed, but the standard is a 185gr (11.998gm) semi-jacketed bullet which develops 1300ft/sec velocity and 695ft/lbs of muzzle energy, which is about 100 per cent improvement over the performance of the .45 ACP.

This policy of taking an existing but elderly design and improving on it is not confined to the Colt M1911. In the last year or two a Hungarian company, FEG, have produced two new pistols based on the well-known 9mm caliber Browning model 1935 'High Power'. First came the FEG Model FP, which was a straight copy of the Browning; this was followed by a 'Competition Model' which had a ventilated rib above the slide, serrated to prevent glare, although the sights were still the standard fixed foresight and laterally-adjustable backsight. These models were later followed by the P9R and P9RA, which were virtually the 'Competition Model' converted to double-action lockwork. The only difference between the P9R and the P9RA is that the latter has the frame made of light alloy and is some 180gm lighter; both have a 118.5mm barrel and a 14-shot magazine, again just like the original Browning. The FEG weapons appear to be well made, and they shoot accurately, but if anyone thinks that because they are copies from behind the Iron Curtain they are going to be cheap, think again; they are little cheaper than the genuine Browning. And since the Browning is now produced in double-action form, there seems little point in going for a substitute unless there is a considerable price advantage.

Whilst these improved pistols are interesting, there is, of course, nothing more interesting to the real enthusiast than a completely new departure in pistol mechanism, and fortunately there have been a few of those in recent years.

Using the propellant gas to delay the opening of a blowback firearm is not entirely new; it was tried in the short-lived 'Volkssturm Gewehr' developed in the later days of World War II as a cheap semi-automatic rifle to arm the German Home Guard, but the war ended before it could be got into production. One or two designers toyed with the idea after

The Colt Python 'Peacekeeper' in .357 Magnum caliber.

that, but it remained for the Steyr-Daimler-Puch factory in Austria to put the idea to use in a practical handgun. In fact they were a little premature with the idea; it first appeared some years ago as the 'Pistole 18' and was licensed to an American manufacturer who had little success with it. Steyr then took it back and overhauled the design, then re-launched it as the 'Model GB', and this time they got it right.

The GB is a man-sized combat pistol, with a magazine capable of taking no less than 18 rounds of 9mm Parabellum without feeling unduly bulky. It is conventional enough in appearance, though the black crackle finish of the frame and matt black finish to the slide give it a distinctive appearance. The trigger is double-action, with an external hammer, and there is a slide-mounted safety and de-cocking lever. After loading a round into the chamber, pressure on this lever locks the firing pin, interposes a steel block between the pin and the hammer, and then drops the hammer, so that the pistol is perfectly safe to carry. When required in a hurry, simply pulling the trigger automatically releases the safety.

The GB is a delayed blowback weapon, and the delay is performed by the propellant gas. The barrel, which is fixed to the frame, is externally shaped to have a piston-head about halfway along; the slide is internally shaped to fit snugly around this piston head and snugly around the muzzle. In front of the piston-head are two gas vents which, when the pistol is fired, allow high-pressure gas to flow into the annular space around the front of the barrel, between the piston-head and the muzzle end of the slide. This high pressure gas resists the rearward movement of the slide by the pressure on the slide front end as the gas is further compressed by the slide's movement. The pressure is sufficient to delay the opening of the breech until the bullet is clear of the muzzle, whereupon the pressure drops because the gas can now be forced back through the vents and into the barrel, to escape to the outside. Moreover the gas action softens the return of the slide and makes the GB a very pleasant pistol to shoot, with soft recoil and relatively little jump, so that it can be brought back on target very quickly.

Most people who see the GB for the first time ask about fouling; surely the gas pouring into the space around the barrel will soon carbon it up and jam the gun? Or even if it doesn't jam the gun, it will be hell to clean? Not so; or rather, not so provided you use good ammunition. Use some old World War II brass which has been lying around for years and there's no telling what you'll get, but use fresh modern ammunition from a reputable maker and there will be no trouble in practical terms. I have seen 300 rounds of Hirten-berg 9mm fired through a GB in the space of an hour or two

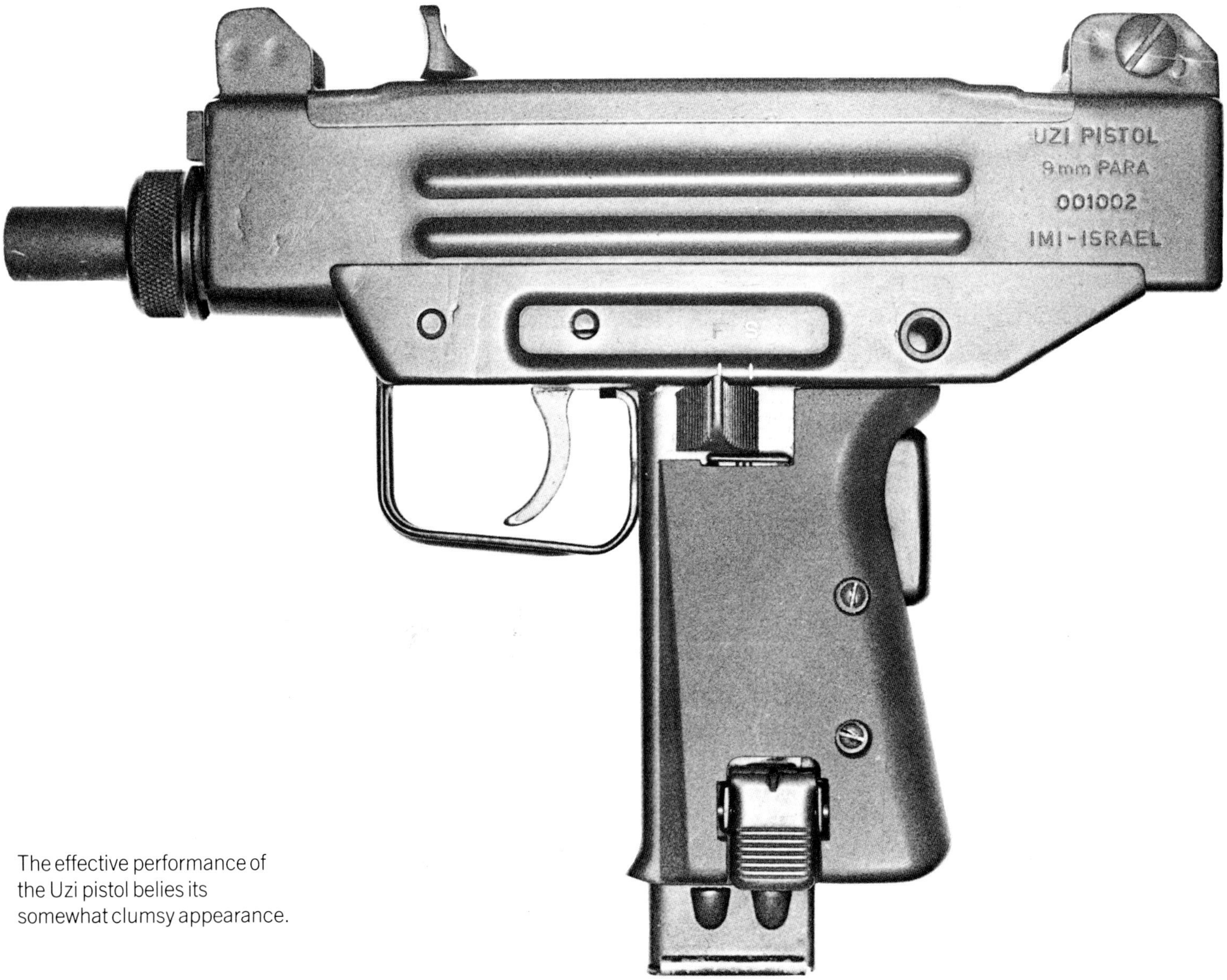

The effective performance of the Uzi pistol belies its somewhat clumsy appearance.

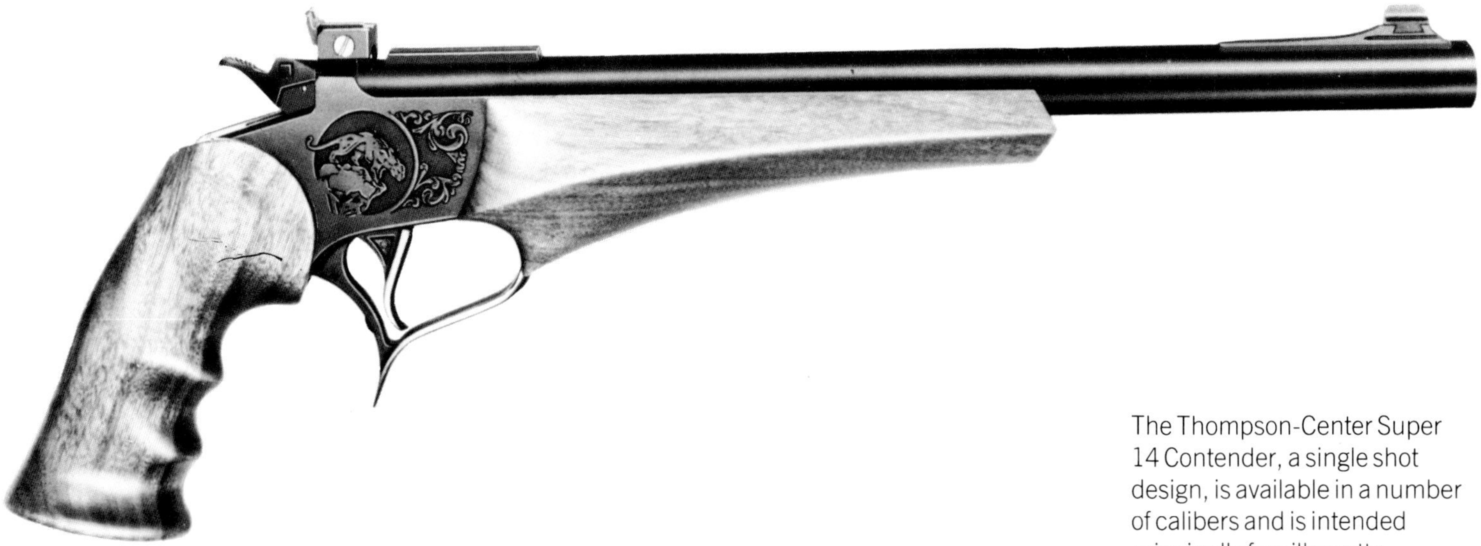

The Thompson-Center Super 14 Contender, a single shot design, is available in a number of calibers and is intended principally for silhouette shooting.

and the stripped pistol showed no signs of fouling other than a dusting of black on the barrel which was immediately removed with an oily rag.

The sights are modified Stavenhagen; a front post and a rear square notch, both with luminous dots for shooting in poor light. The front sight has a single dot, the rear sight has two dots alongside the notch; line all three dots up and the aim is correct for 25 meters range. And the accuracy is as good as the man holding the pistol; at a demonstration in the USA some months ago a Steyr demonstrator put 18 shots through a man-sized target 175 meters away, and used tracer ammunition so that the audience could see that there was no trickery.

Another design with something different to offer is the Italian Benelli B-76. Once again, this is a full-size pistol aimed at the military and police market, and it is available in the usual 9mm Parabellum caliber.

The B-76 looks quite conventional, a double-action automatic with an external hammer and a frame-mounted safety catch, but the novelty is concealed under the slide. It is one of the rare class of weapon using a toggle system to lock the breech during firing. There is a separate breech-block carried in the rear of the slide and the two are connected by a dumb-bell shaped toggle. As the slide runs forward, after loading, the breech block comes to a stop against the cartridge, but the slide continues to move until it has moved the toggle joint over-center so that it forces the rear end of the breech-block downwards to lock into a recess in the frame and the toggle forms a rigid strut. On firing the pressure on the cartridge case forces the breech-block harder into the locking recess, so the breech cannot open. But at the same time pressure is placed on the toggle and, due to its mechanical arrangement, this takes some time to move and lift the rear of the block free of the lock. By the time the block lifts, the bullet is clear of the muzzle and the chamber pressure has dropped to a safe level. The toggle now transfers thrust from the breech-block to the slide, and both move back together, eject the spent case and begin the loading cycle over again.

On the other hand if your tastes run to something aggressively military and quite distinctive, then the Uzi pistol is for you. Anyone seeing it for the first time will immdiately think that it is a shrunken Uzi submachine gun, but it is even shorter than the 'Mini-Uzi' and has been engineered so as to fire only in the semi-automatic mode and so that it cannot be 'reverse engineered' to turn it into a full-auto weapon. It is exactly the same shape as the well-known submachine gun but smaller in all dimensions except caliber, since it fires the same 9mm Parabellum cartridge. Nine inches overall, it looks clumsy and cumbersome, but in fact the weight (just over 4lbs loaded with a 20 round magazine) means that it sits firmly in the hand, doesn't buck very much when fired and is easy to control when firing a rapid succession of shots. The size and shape also makes it easy to hold two-handed. Like the submachine gun it has a cocking handle on the top of the receiver, a grip safety in the rear of the pistol grip, and a thumb-operated safety catch on the left side of the frame.

No pistol is of much use without ammunition, and there have been some interesting moves in ammunition design over the past two or three years. You might think that there is a sufficiency of calibers and cartridges already, without having to add to the list, but there is always room for something new if it promises to deliver something that the others don't have.

For revolver shooters there are two cartridges of considerable interest. Firstly, for those demanding power and plenty of it, there is the .357 Remington Maximum. This powerful cartridge was introduced in 1982 by a collaboration between Remington and Sturm Ruger, and a Ruger revolver was provided for it. Since then a Dan Wesson revolver and a single-shot pistol by Thompson/Center have also been chambered for it. The cartridge is almost identical to the .357 Magnum except that the case is 8mm longer and carries a heavier propelling charge. The standard Remington loading is a 10.24gm (158gr) semi-jacketed hollow point bullet, and the Federal Cartridge Co. produced an 11.66gm (180gr) bullet of similar type. The Remington factory cartridge will deliver 1824ft/sec in a 266mm (10.5in) test barrel, and this produces a muzzle energy figure of 1170ft/lbs; performance in actual pistols will be slightly different due to different barrel lengths. Nevertheless, this is a vast improvement over the standard .357 Magnum figures of 1430ft/sec and 720ft/lbs in a 6-inch barrel.

The .357 Remington Maximum was developed primarily for long range metallic silhouette shooting. Tests show that it is a very accurate cartridge, ideally suited for this form of competition, but unfortunately it is a very hot loading and it has caused problems with barrel erosion and with flame-cutting of the top strap of the revolver, above the cylinder-barrel junction. Some more research is underway on this, probably with a view to altering the propellant composition so as to cut down on the erosion, and there are signs that other revolver manufacturers intend to produce suitably chambered weapons.

The second revolver cartridge to be recently introduced is the .32 Harrington & Richardson Magnum. To speak of 'Magnum' in .32 calibre may sound a little odd, but there is some good sound sense behind the idea, as might be imagined when a company as old and respected as H&R put their name behind it. The .32 H&R Magnum is, as you might guess, a lengthened .32 case with a more powerful powder charge inside it, carrying a 6.16gm (95gr) semi-wadcutter bullet. According to the manufacturers, 'the performance of the .32 Magnum exceeds that of the .38 Special service loads.' There are, though, innumerable 'service' loads for the .38 Special and some of them are certainly more potent than this .32 cartridge. Nevertheless, it does give a useful increase over the normal performance, delivering 1030ft/sec velocity and 225ft/lbs muzzle energy; a comparable bullet in .32 S&W Long caliber would deliver about 780ft/sec and 132ft/lbs.

As to the purpose behind it; well, for many years the .32 was virtually a standard police caliber in the USA, and there are many people who would be glad of the reduced size and better concealability of a .32 revolver if it could be endowed with sufficient power to make it a reasonable self-defence cartridge. In the .32 H&R Magnum they have now got that cartridge, though it must be pointed out that it is not intended to be used in existing .32 revolvers, since it is too powerful. As it happens the lengthened case will prevent chambering the cartridge in any unsuitable .32, and H&R have produced revolvers with suitably lengthened and strengthened cylinders. It is to be expected that other manufacturers will follow suit, and it looks as if the .32 is in for something of a revival. With the semi-wadcutter bullet it would make a nice target weapon, and it would also be useful as a kit gun for campers and outdoorsmen to pack along.

The automatic pistol shooter has two new cartridges to think about, the 9mm Winchester Magnum Auto and the .45

The 9mm Mamba, a robust and accurate South African design.

Winchester Magnum Auto. Both cartridges were developed by Winchester in conjunction with Wildey Arms in 1979 to suit the Wildey gas operated automatic pistol, and there have since been indications that one or two other pistol manufacturers may chamber more conventional pistols to suit them. The .45 Magnum Auto is more or less the standard .45 ACP case lengthened by some 7mm and fitted with one of the heavier .45 ACP bullets, a full metal jacketed type weighing 14.9gm (230gr). The charge is, of course, more powerful than that commonly used with the .45 ACP and thus the performance is quite impressive; in a five-inch barrel it develops 1400ft/sec and 1005ft/lbs, whereas a .45 ACP with the same bullet will just make 860ft/sec and 425ft/lbs. Once again the round has been developed with a metallic silhouette shooting in mind, and, frankly, there seems to be little else it could be used for. We gather that some military agencies looked at it and decided that it was far too powerful to be used in a service handgun, if only from the point of view that it would be difficult to get the average soldier to use it efficiently. But it does seem that there might be a future in some form of semi-automatic carbine.

At the same time as the .45 was developed, the same people produced the 9mm Magnum Auto, again for a Wildey pistol and again for silhouette shooting. The case is some 5mm longer than a standard 9mm Parabellum, and it is loaded with a 7.45gm (115gr) full metal jacketed bullet to give a muzzle velocity of 1475ft/sec and a muzzle energy of 558ft/lbs. By comparison a Winchester 9mm Parabellum using the same bullet produces 1225ft/sec and 383ft/lbs of energy, so the Magnum Auto gives a useful improvement.

But on closer examination one is entitled to ask a question; what does this offer which wasn't previously offered by the venerable 9mm Mauser Export cartridge? That cartridge, developed as long ago as 1908, used an 8.15gm (126gr) bullet to produce 1360ft/sec and 520ft/lbs, and presumably with the lighter 115gr bullet the performance would be so much closer to that of the 9mm Magnum Auto that the difference would be negligible. Admittedly, the 9mm Mauser doesn't seem to have been made anywhere since 1945, but it has a good record as a submachine gun cartridge, which argues that there might be a military niche for the 9mm Magnum Auto one day.

Mention of the 9mm Mauser Export calls to mind another old but good cartridge which appears to be enjoying a revival; the .38 Super Auto. This first appeared as the .38 Auto in 1900, to suit a Colt pistol, and was hotted up to .38 Super in about 1929. For many years it was the most potent US automatic pistol cartridge, and even the Thompson submachine gun was chambered for it. It has always been a popular chambering in a number of Spanish pistols, and recently the Auto-Ordnance Company, the firm who developed and produced the Thompson, have ventured into the pistol field with an automatic based on the M1911 design and chambered for, among others, the .38 Super cartridge. With a 130gr full metal jacketed bullet the .38 Super will deliver 1280ft/sec and 474ft/lbs.

Probably the most peculiar cartridge to appear in recent months is a new Soviet military loading, the 5.45mm Soviet Pistol round. This has been known in the West since 1979 in outline form, but it was not until 1983 that more details, together with some information on the parent weapon, were discovered. The pistol is the 'PSM' (Pistolet Samozaryadniy Malogabaritny – pistol, self-loading, small) which is currently being issued to Soviet security troops and police. A few specimens have also been seen in Cuba and Central America. It is a simple blowback weapon, resembling the Walther PPK in size and appearance and is double-action with an external hammer. The major mechanical change which is visible is that the safety has been recessed into the end of the slide, rather than on the side, so as to make the pistol as thin as possible to aid in concealment. Total length is 6.3 inches, and with a full 8 shot magazine it weighs 17 ounces.

The unusual Soviet PSM 5.45mm automatic.

The 5.45mm cartridge for this pistol is a peculiar rimless bottle-necked round weighing 77gr and with a case length of 0.98 inch. The bullet is full metal jacketed, flat-tipped, with a compound core and weighs 40gr. The most remarkable thing about this bullet is its length; it is 0.57 inches long, which is 2.58 calibers; the normal caliber-length for pistol bullets is around 1.5. The next remarkable thing is its construction, since it has a gilding-metal jacket surrounding a core which has the front half made of steel and the rear half of lead. This is a most unusual form of construction for a pistol bullet, and it closely resembles the 5.45mm rifle bullet developed for the Soviet AK-74 assault rifle, leading us to suppose that there is some degree of commonality in the manufacturing machinery and process for the sake of economy.

Tests carried out on this bullet have shown some odd things. In the first place, there isn't much stopping power if we base our considerations on arithmetic; the bullet develops 1033ft/sec muzzle velocity and 174ft/lbs of energy, which is almost identical to the power of the .25 ACP cartridge, which few people take seriously. But the shape and balance of the bullet appear to give it a very good accuracy, and, most remarkable of all, it has enormous penetrative power for its class; at ten yards range it will go through a GI helmet just as easily as a 5.56mm rifle bullet, and that's something the .25 ACP won't do.

The biggest puzzle is why the Soviets adopted a bottle-necked cartridge case; bottle-necked cases are far from being the optimum choice in a blowback weapon, though in such a low-power pistol as this they are unlikely to give much trouble in ejecting. But there are half a dozen existing commercial cartridges, plus a couple of existing Soviet rounds, which would have been far easier to design the pistol around and which would have given equal, if not better, performance. It is a well-known fact that the Soviets never adopt a cartridge capable of being fitted into a Western weapon, but they seem to have over-reached themselves this time.

SPECIFICATIONS OF REPRESENTATIVE MODERN PISTOLS

NAME	CALIBER	OPERATION	LENGTH	WEIGHT EMPTY	BARREL LENGTH	MAGAZINE CAPACITY	REMARKS
AGNER: Agner-Saxhoj Products, Saxhoj, Denmark							
Model 80	.22LR	Blowback	10.10in	42.0oz	6.00in	5	Match pistol
AMT: Arcadia Machine & Tool Co., Covina CA							
Hardballer	.45 ACP	Recoil	8.50in	39.0oz	5.00in	7	
Lightning	.22LR	Blowback	10.75in	38.0oz	6.50in	10	Also 8.5in, 10.5in and 12.5in barrels
ARMINEX: Arminex Ltd., Scottsdale, AZ 85258							
Trifire	.45ACP	Recoil	8.5in	38.0oz	5.00in	9	Also .38 Super and 9mm Para.

The Arminex Trifire, one of the many improved versions of the famous M1911.

NAME	CALIBER	OPERATION	LENGTH	WEIGHT EMPTY	BARREL LENGTH	MAGAZINE CAPACITY	REMARKS
ARMINIUS: Hermann Weirauch KG, D-8744 Mellrichstadt/Bayern, West Germany							
HW3	.22LR	Revolver	7.00in	26.4oz	2.80in	8	Also .32S&W
HW5	.22LR	Revolver	8.70in	27.5oz	4.00in	8	Also .22WMRF and .32 S&W
HW7	.22LR	Revolver	10.40in	30.9oz	6.00in	8	Also .22WMRF or .32 S&W
HW357	.357M	Revolver	9.10in	31.7oz	4.00in	6	Also 2.5in and 6.0in barrels
HW38	.38Spl	Revolver	8.90in	30.9oz	4.00in	6	Also 2.5in and 6.0in barrels
HW68	.22LR	Revolver	6.70in	18.7oz	2.50in	8	Also .32S&W

ASP: Armament Systems & Procedures Inc, Box 356, Appleton, WI 54912

NAME	CALIBER	OPERATION	LENGTH	WEIGHT EMPTY	BARREL LENGTH	MAGAZINE CAPACITY	REMARKS
Asp	9mmP	Recoil	7.40in	20.1oz	2.24in	7	Conversion of S&W M39

The ASP 9mm pistol is based on the Smith & Wesson Model 39. Modifications include the trough sight pictured separately and the transparent panel on the left side of the magazine.

ASTRA: Astra Unceta y Cia SA, Apartado 3, Guernica, Vizcaya, Spain

NAME	CALIBER	OPERATION	LENGTH	WEIGHT EMPTY	BARREL LENGTH	MAGAZINE CAPACITY	REMARKS
Model A-80	9mmP	Recoil	7.08in	34.7oz	3.80in	15	Also in .45ACP, 9mm Steyr, .38 Super Auto, 7.65mm Para.
Model A-50	7.65mm	Blowback	6.61in	23.2oz	3.50in	8	Also 9mm/.380
Falcon	9mm/.380	Blowback	6.45in	22.7oz	3.88in	7	Also 7.65mm
Model 960	.38Spl	Revolver	9.46in	40.5oz	4.01in	6	Also 6in bbl.
Model 41	.41Mag	Revolver	11.54in	45.5oz	6.0in	6	Also 8.5in bbl.
Model 44	.44Mag	Revolver	11.54in	45.1oz	6.0in	6	Also 8.5in bbl.
Model 45	.45 Colt	Revolver	11.54in	43.7oz	6.0in	6	Also 8.5in bbl.
Police	.357Mag	Revolver	8.35in	36.7oz	3.03in	6	

The Astra .44 Magnum revolver. Some shooters find the grip on this gun somewhat large.

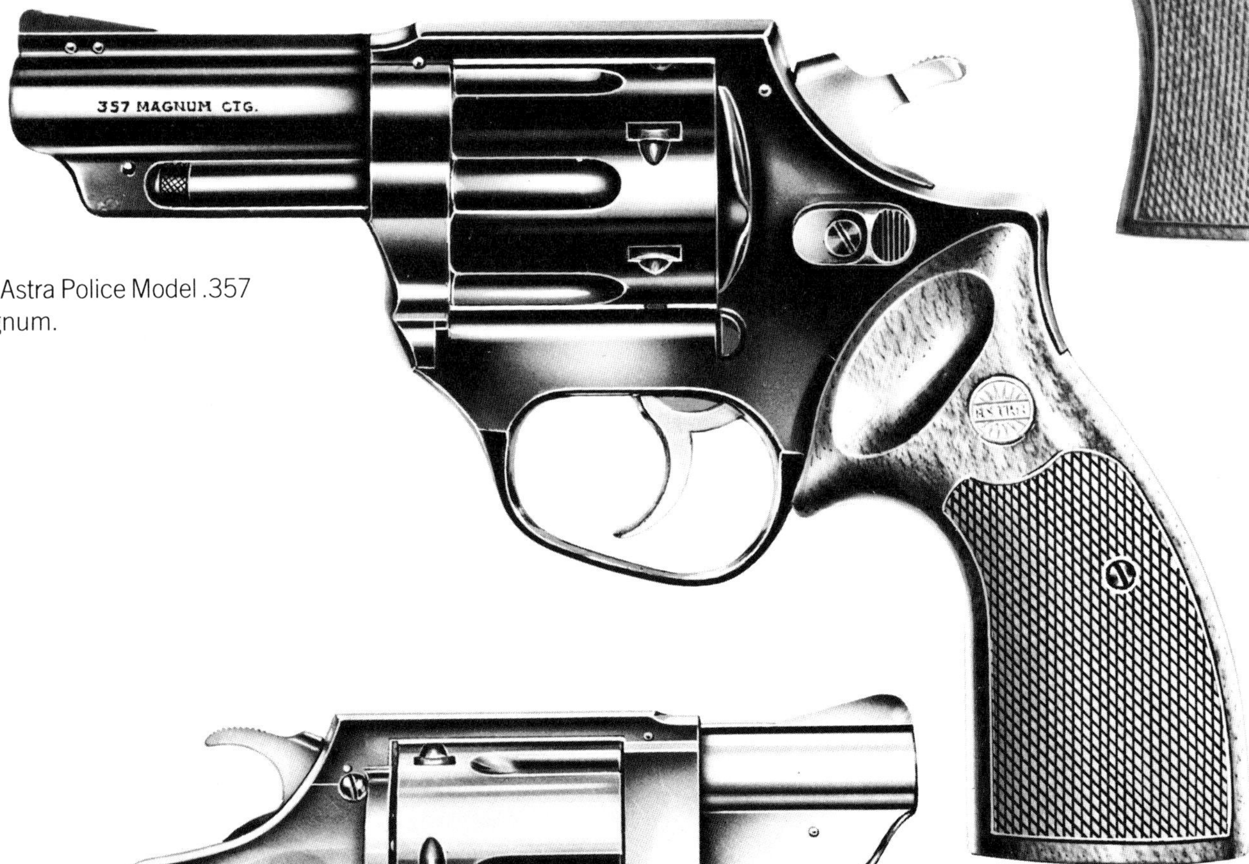

The Astra Police Model .357 Magnum.

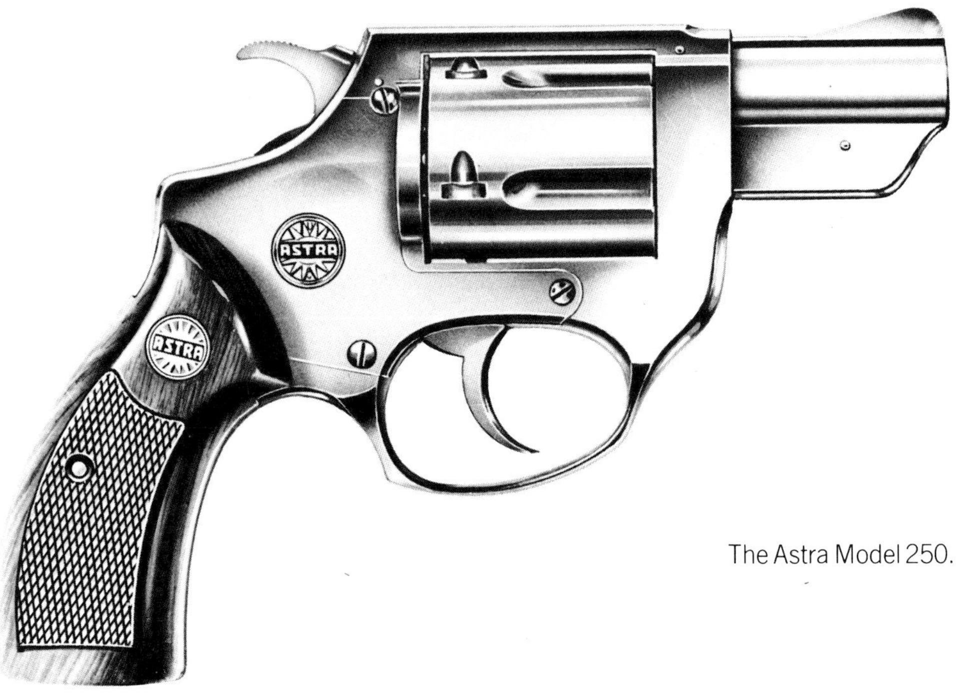

The Astra Model 250.

NAME	CALIBER	OPERATION	LENGTH	WEIGHT EMPTY	BARREL LENGTH	MAGAZINE CAPACITY	REMARKS
ASTRA Contd.							
Constable	7.65mm	Blowback	6.60in	22.8oz	3.50in	8	Also 9mm/.380 or .22LR
Cub	6.35mm	Blowback	4.40in	12.4oz	2.30in	7	Also sold in US as Colt Junior
Match	.38Spl	Revolver	11.30in	43.2oz	6.00in	6	Target pistol
Cadix	.38Spl	Revolver	9.00in	28.8oz	4.00in	6	Also .22 and .32
AUTO NINE: Auto Nine Corp., Parma ID 83660							
Auto Nine	.22LR	Blowback	4.31in	7.0oz	2.15in	8	
BAUER: Bauer Firearms Corp., Fraser, MI 48026							
Auto	6.35mm	Blowback	4.00in	10.0oz	2.30in	6	
BENELLI: Benelli Armi SpA, Urbino, Italy							
Model B76	9mmP	Recoil	8.10in	34.2oz	4.30in	8	Toggle lock

The Benelli Model B76 is the first military-style pistol from this maker, previously better known for their sporting weapons.

NAME	CALIBER	OPERATION	LENGTH	WEIGHT EMPTY	BARREL LENGTH	MAGAZINE CAPACITY	REMARKS
BERETTA: Pietro Beretta SpA, Gardone Val Trompia, Brescia, Italy							
Model 20	6.35mm	Blowback	4.90in	10.9oz	2.40in	8	Double action
Model 70	7.65mm	Blowback	6.50in	23.3oz	3.50in	8	Updated M1934
Model 76	.22LR	Blowback	9.20in	32.8oz	5.90in	10	Match pistol
Model 81	7.65mm	Blowback	6.77in	23.5oz	3.82in	12	Double action
Model 81BB	7.65mm	Blowback	6.77in	23.5oz	3.82in	12	Improved 81
Model 82BB	7.65mm	Blowback	6.77in	22.2oz	3.82in	9	Improved 81
Model 84	9mm/.380	Blowback	6.77in	22.6oz	3.82in	13	Double action

NAME	CALIBER	OPERATION	LENGTH	WEIGHT EMPTY	BARREL LENGTH	MAGAZINE CAPACITY	REMARKS
BERETTA Contd.							
Model 84BB	9mm/.380	Blowback	6.77in	22.6oz	3.82in	13	Improved 84
Model 85BB	9mm/.380	Blowback	6.77in	21.9oz	3.82in	8	
Model 87	.22RF	Blowback	6.77in	20.1oz	3.82in	10	
Model 92	9mmP	Recoil	8.54in	33.5oz	4.92in	15	Double action
Model 92S	9mmP	Recoil	8.54in	34.6oz	4.92in	15	Modified 92
Model 92SB	9mmP	Recoil	8.54in	34.6oz	4.92in	15	Improved 92S
Model 92SB-C	9mmP	Recoil	7.75in	32.4oz	4.29in	13	Compact 92SB
Model 92SB-C Type M	9mmP	Recoil	7.75in	31.9oz	4.29in	8	Small magazine
Model 93R	9mmP	Recoil	9.45in	39.5oz	6.14in	20	Selective auto

The Beretta Model 93R with forward hand grip folded down and the selector set for three-round bursts. The shoulder stock can also be fitted.

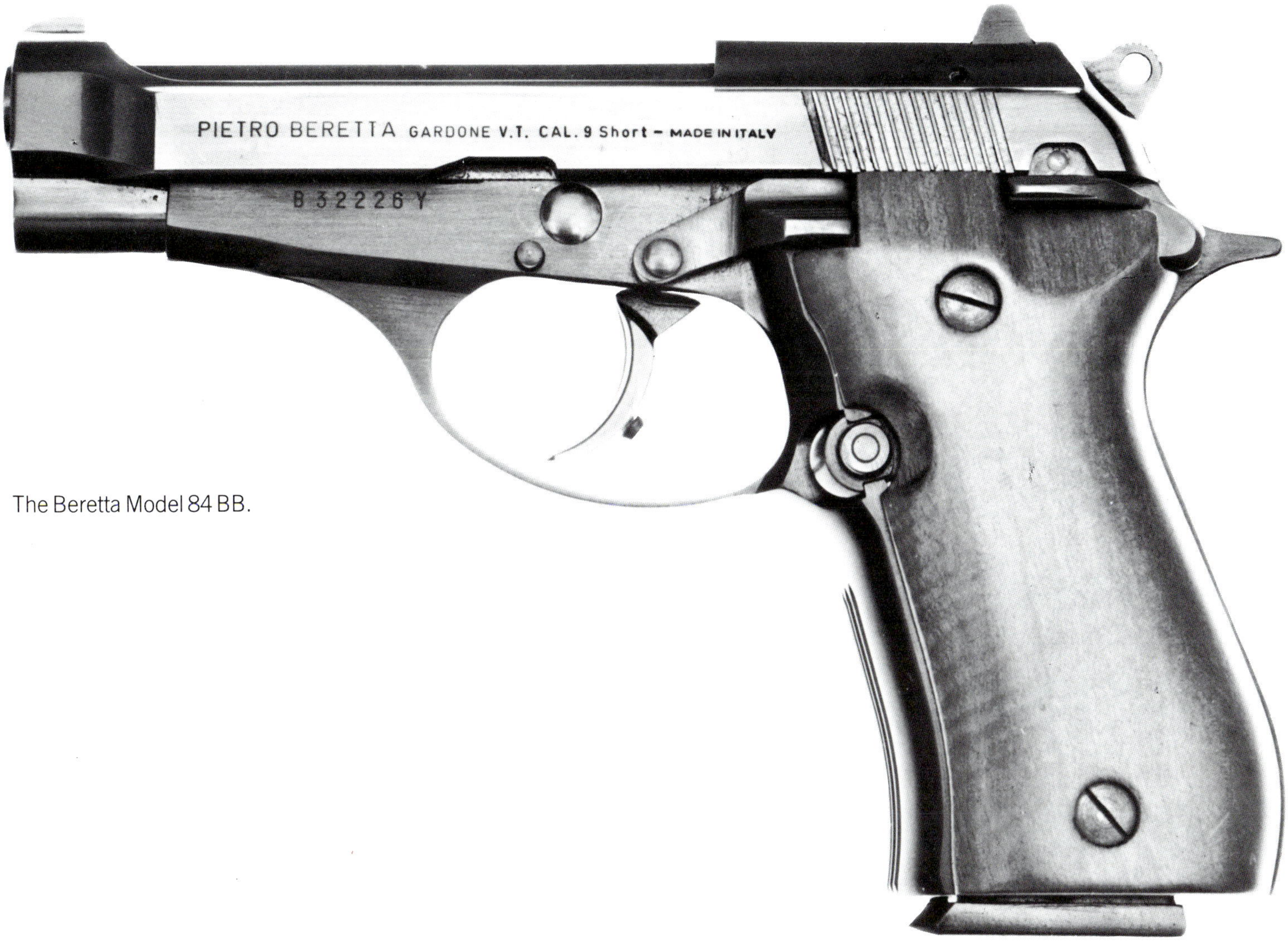

The Beretta Model 84 BB.

The Beretta Model 92 showing the double-column magazine. This weapon was first produced in 1976 and is used by the Italian forces.

NAME	CALIBER	OPERATION	LENGTH	WEIGHT EMPTY	BARREL LENGTH	MAGAZINE CAPACITY	REMARKS
BERETTA Contd.							
Model 98	7.65mmP	Recoil	7.75in	31.7oz	4.29in	13	New caliber SB-C
Model 99	7.65mmP	Recoil	7.75in	31.5oz	4.29in	8	New caliber SBC Type M
Model 951R	9mmP	Recoil	6.69in	47.6oz	4.92in	10	Selective auto
Model 952 Special	7.65mmP	Recoil	9.80in	33.8oz	5.90in	8	Target pistol

The Beretta Model 951R machine pistol.

BERNADELLI: Vincenzo Bernadelli SpA, I-25063, Gardone Val Trompia, Brescia, Italy

Model 60	7.65mm	Blowback	6.50in	23.8oz	3.50in	8	Also .22LR and 9mm/.380
Model 68	6.35mm	Blowback	4.30in	9.5oz	2.25in	5	
Model 69	.22LR	Blowback	9.00in	38.0oz	5.90in	10	Target pistol
P-018	9mmP	Recoil	8.39in	35.2oz	4.80in	15	Double action
Model VB	.32 S&W	Revolver	7.60in	18.3oz	3.10in	6	Also .22LR and .22 WMRF. In other barrel lengths

NAME	CALIBER	OPERATION	LENGTH	WEIGHT EMPTY	BARREL LENGTH	MAGAZINE CAPACITY	REMARKS
BROWNING: Fabrique Nationale, B-4400 Herstal, Belgium							
Model 125	9mm/.380	Blowback	7.00in	25.4oz	3.80in	9	
Model DA140	7.65mm	Blowback	6.81in	22.5oz	3.80in	12	Also 9mm Short
High-Power	9mmP	Recoil	7.87in	31.1oz	4.64in	13	
High-Power Mk 2	9mmP	Recoil	7.87in	31.1oz	4.64in	13	Ambidextrous safety catch
FN-DA	9mmP	Recoil	7.87in	32.4oz	4.64in	14	Double action
FN-DAM	9mmP	Recoil	7.00in	30.7oz	3.78in	14	Double action
FN-DAC	9mmP	Recoil	6.81in	26.8oz	3.78in	7	Double action
Barracuda	.357M	Revolver		37.0oz	3.00in	6	Also .38Spl and 9mmP
Challenger	.22LR	Blowback	8.90in	35.0oz	6.60in	10	Target pistol
International	.22LR	Blowback	11.70in	46.7oz	5.90in	10	Target pistol

The Browning High Power double action automatic.

NAME	CALIBER	OPERATION	LENGTH	WEIGHT EMPTY	BARREL LENGTH	MAGAZINE CAPACITY	REMARKS
CENTRUM: VEB Ernst Thalmann, DDR-60 Suhl, East Germany							
Centrum	.22LR	Single shot	13.90in	45.0oz	10.00in	–	Match pistol
Model IV	.22LR	Blowback	11.80in	47.7oz	6.00in	5	Match pistol

CHARTER ARMS: Charter Arms Corp., Stratford, CT 06497

NAME	CALIBER	OPERATION	LENGTH	WEIGHT EMPTY	BARREL LENGTH	MAGAZINE CAPACITY	REMARKS
Bulldog	.357M	Revolver	7.30in	19.0oz	2.50in	5	Also .44Spl
Bulldog Tracker	.357M	Revolver	9.00in	24.0oz	4.00in	5	Also 4in and 6in barrels
Explorer 2	.22LR	Blowback	15.5in	28.0oz	8.00in	10	Also 6in and 10in barrels
Pathfinder	.22LR	Revolver	6.30in	19.0oz	2.00in	6	Also 3in and 6in barrels
Police Bulldog	.38Spl	Revolver	7.10in	19.0oz	2.00in	5	Also 4in bbl
Target Bulldog	.357M	Revolver	9.00in	29.0oz	4.00in	5	Also .44Spl
Undercover	.38Spl	Revolver	6.30in	16.0oz	2.00in	5	Also .32S&W and with 3in bbl

The Charter Arms Target Bulldog in .44 Special.

COLT: Colt Industries, Hartford, CT 06102

NAME	CALIBER	OPERATION	LENGTH	WEIGHT EMPTY	BARREL LENGTH	MAGAZINE CAPACITY	REMARKS
Ace	.22LR	Blowback	8.40in	42.0oz	4.70in	10	
Combat Commander	.45ACP	Recoil	7.90in	36.0oz	4.70in	8	Also .38 Super and 9mm Para

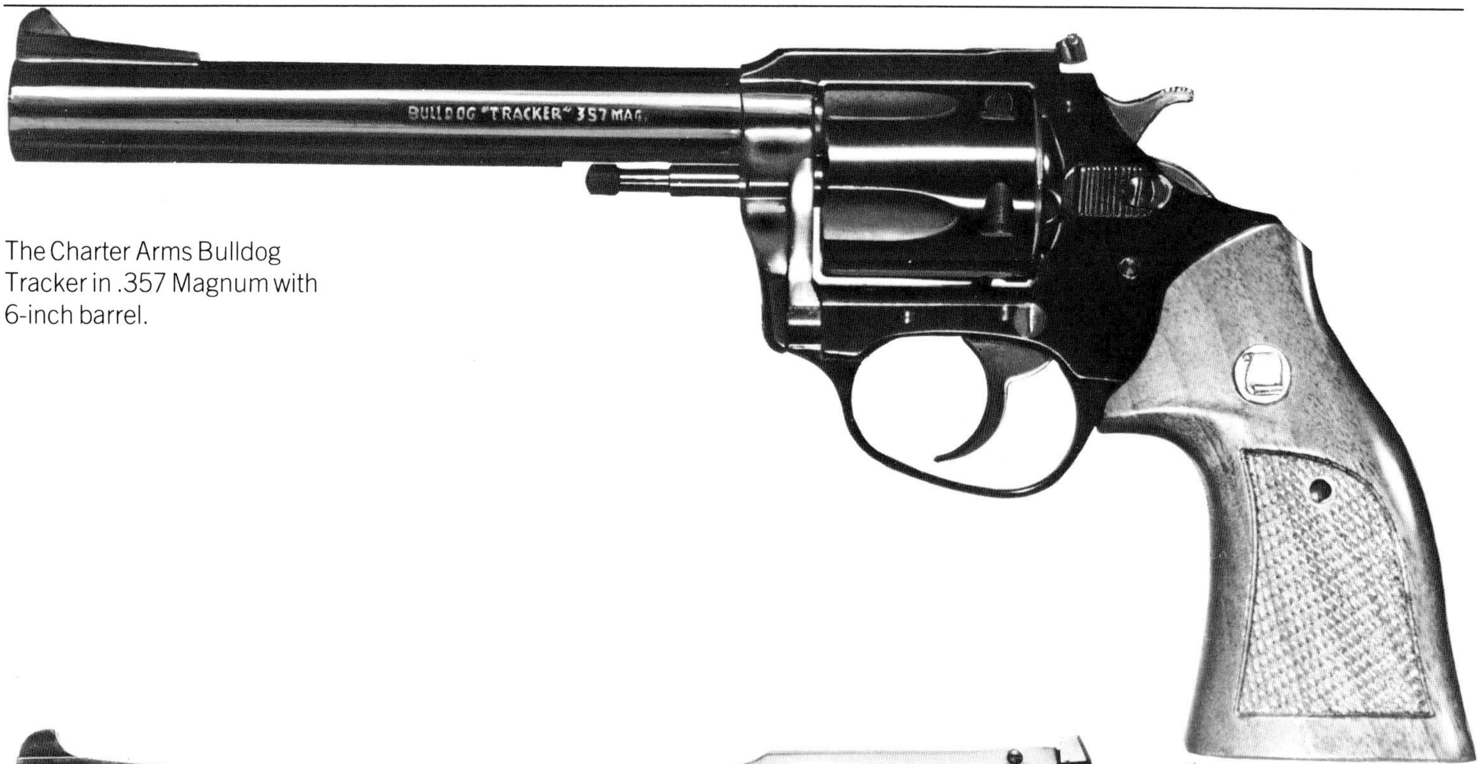

The Charter Arms Bulldog Tracker in .357 Magnum with 6-inch barrel.

The Charter Arms Pathfinder in standard form.

The Colt Series 80, a slightly more compact version of the famous M1911.

NAME	CALIBER	OPERATION	LENGTH	WEIGHT EMPTY	BARREL LENGTH	MAGAZINE CAPACITY	REMARKS
COLT Contd.							
Detective Special	.38Spl	Revolver	6.90in	21.5oz	2.00in	6	
Diamondback	.38Spl	Revolver	9.00in	31.8oz	4.00in	6	Also 6in bbl
Government Model	.45ACP	Recoil	8.40in	38.0oz	4.70in	8	Also .38 Super and 9mm Para
Lawman Mk 3	.357M	Revolver	7.40in	32.0oz	2.00in	6	Also 4in bbl
New Frontier	.22LR	Revolver	9.80in	31.0oz	4.40in	6	Single action. Also 6 or 7.5in barrels
Python	.357M	Revolver	11.30in	43.5oz	6.00in	6	Also 2.5in, 4in and 8in barrels. And in .38Spl with 8in barrel
Trooper Mk 5	.357M	Revolver	11.50in	46.0oz	6.00in	6	Also 4in and 8in barrels; also in .22RF or .22WMRF
Series 80	.45ACP	Recoil	7.25in	34.0oz	3.63in	6	Compact 1911

The Colt Agent .38 Special.

NAME	CALIBER	OPERATION	LENGTH	WEIGHT EMPTY	BARREL LENGTH	MAGAZINE CAPACITY	REMARKS
CZ: Ceskoslovenska Zbrojovka, Uhersky Brod, Czechoslovakia							
CZ75	9mmP	Recoil	8.00in	34.5oz	4.72in	15	Double action
CZ83	7.65mm	Blowback	6.81in	22.9oz	3.78in	15	Double action. Also 9mm Short
DAN WESSON: Dan Wesson Co., Monson, MA 01057							
44 Magnum	.44Mag	Revolver	10.00in	45.0oz	4.00in	6	Also 6, 8, 10in barrels
Model 15-2	.357M	Revolver	9.25in	36.0oz	4.00in	6	Also 2, 6, 8, 10, 12 and 15in barrels
DETONICS: Detonics .45 Associates, Seattle Tower Building, Seattle WA 98101							
Combat Master	.45ACP	Recoil	6.73in	26.0oz	3.5in	5	
Scoremaster	.45ACP	Recoil	8.75in	41.0oz	5.0in	7	Also .451 cal
Pocket 9	9mmP	Recoil	7.88in	26.0oz		6	

The Detonics Pocket 9. The Detonics company is best known for its compact pistol designs.

NAME	CALIBER	OPERATION	LENGTH	WEIGHT EMPTY	BARREL LENGTH	MAGAZINE CAPACITY	REMARKS
DRULOV: Dino Svratovch, Litomysl, Czechoslovakia							
Model 70	.22LR	Single shot	14.40in	44.1oz	9.80in	–	Target pistol
Model 75	.22LR	Single shot	14.40in	44.1oz	9.80in	–	Modified 70
Pav	.22LR	Single shot	10.20in	22.9oz	7.90in	–	Target pistol

NAME	CALIBER	OPERATION	LENGTH	WEIGHT EMPTY	BARREL LENGTH	MAGAZINE CAPACITY	REMARKS
ERMA: Erma-Werke GmbH, Postbox 1269, D-8060 Munich-Dachau, West Germany							
KGP-69	.22LR	Blowback	7.40in	29.6oz	4.00in	8	Luger copy
ER-422	.22LR	Revolver	7.30in	21.2oz	3.00in	6	
ER-438	.38Spl	Revolver	7.30in	20.1oz	3.00in	5	
EP-552	.22LR	Blowback	5.40in	14.5oz	2.90in	7	
F.A.S.: Fabbrica Armi Sportive SrL, I-20091 Settimo Milano, Italy							
Model OP-601	.22Short	Blowback	11.30in	27.2oz	6.00in	5	Match pistol
Model OP-602	.22LR	Blowback	11.20in	28.6oz	5.60in	5	Sport pistol
Model CF-603	.32S&W	Blowback	11.20in	42.3oz	5.30in	5	Target pistol
F.E.G.: Femaru es Gepgyar NV, Budapest, Hungary							
Model R	7.65mm	Blowback	5.52in	15.9oz	2.83in	6	Chromed bore
Model FP	9mmP	Recoil	7.87in	31.1oz	4.64in	13	Copy of Browning High Power
Model P9R	9mmP	Recoil	7.99in	35.3oz	4.67in	14	Double action
Model P9RA	9mmP	Recoil	7.99in	28.9oz	4.67in	14	Light alloy frame

The F. E. G. Model P9RA double action design.

NAME	CALIBER	OPERATION	LENGTH	WEIGHT EMPTY	BARREL LENGTH	MAGAZINE CAPACITY	REMARKS
HAMMERLI: Hammerli AG, CH-5600 Lenzburg, Switzerland							
Model 120	.22LR	Single shot	15.70in	43.4oz	10.00in	–	Free pistol
Model 150	.22LR	Single shot	16.90in	45.8oz	10.00in	–	Free pistol
Model 152 Electronic	.22LR	Single shot	16.90in	46.9oz	10.00in	–	Free pistol
Model 208	.22LR	Blowback	10.00in	34.7oz	5.90in	8	Target pistol
Model 212	.22LR	Blowback	8.50in	31.0oz	5.00in	8	Sport pistol
Model 230	.22Short	Blowback	11.60in	43.8oz	6.30in	5	Target pistol
HARRINGTON & RICHARDSON: Harrington & Richardson, Gardner MA 01440							
Model 504	.32H&RMag	Revolver	10.75in	33.0oz	6.00in	5	Also 4in bbl
Model 603	.22WMRF	Revolver		35.0oz	6.00in	6	
Model 622	.22LR	Revolver		26.0oz	4.00in	6	
Model 649	.22LR	Revolver		31.0oz	7.50in	9	Also .22WMRF
Model 732	.32S&W	Revolver		26.0oz	4.00in	6	
Model 826	.22WMRF	Revolver		28.0oz	3.00in	6	Bull barrel
Model 999	.22LR	Revolver		34.0oz	6.00in	9	

The Hammerli P240 target pistol, like the company's other designs, has an enviable reputation for accuracy.

HECKLER & KOCH: Heckler & Koch GmbH, D-7238 Oberndorf-am-Neckar, West Germany

NAME	CALIBER	OPERATION	LENGTH	WEIGHT EMPTY	BARREL LENGTH	MAGAZINE CAPACITY	REMARKS
HK4	7.65mm	Blowback	6.20in	16.9oz	3.30in	8	Also .22LR, 6.35mm and 9mm/.380
P9S	9mmP	Delayed Blowback	7.56in	31.0oz	4.01in	9	Also .45ACP
VP70M	9mmP	Blowback	8.03in	29.0oz	4.56in	18	Double action and burst fire. VP70Z without burst fire option
P7	9mmP	Blowback	6.73in	27.5oz	4.13in	8 or 13	Recoil brake

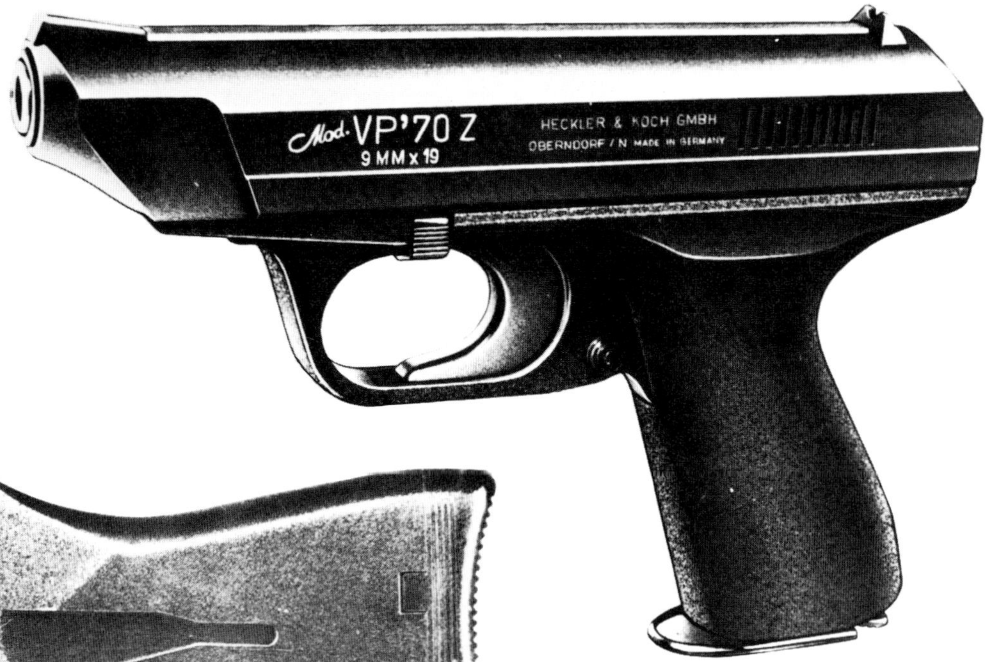

Right: The Heckler & Koch VP70Z, the civil version of their military VP70M pistol.

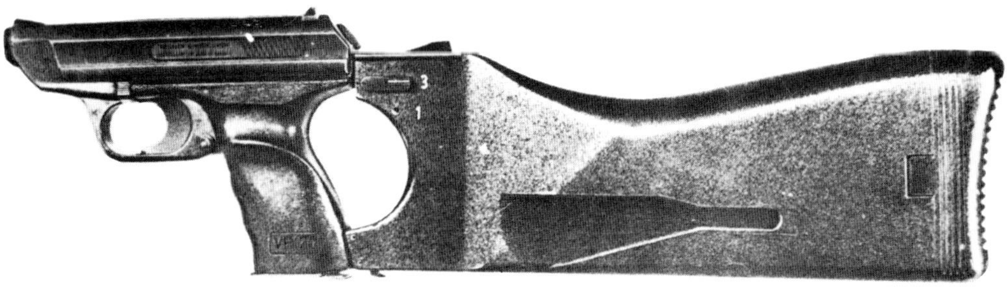

Below: The VP70M with the shoulder stock fitted, an action which automatically engages the three round burst mechanism.

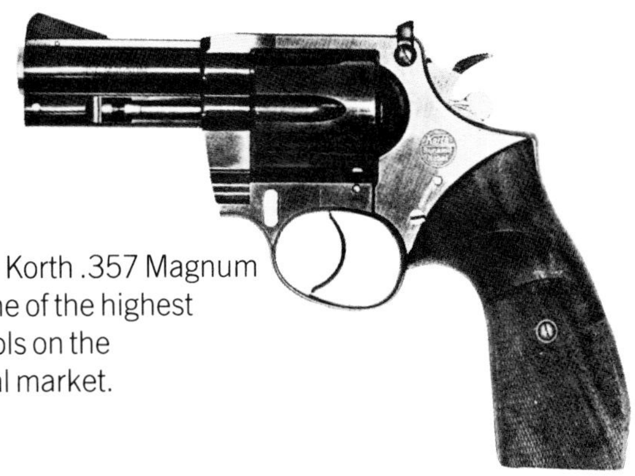

Above: The Korth .357 Magnum revolver, one of the highest priced pistols on the commercial market.

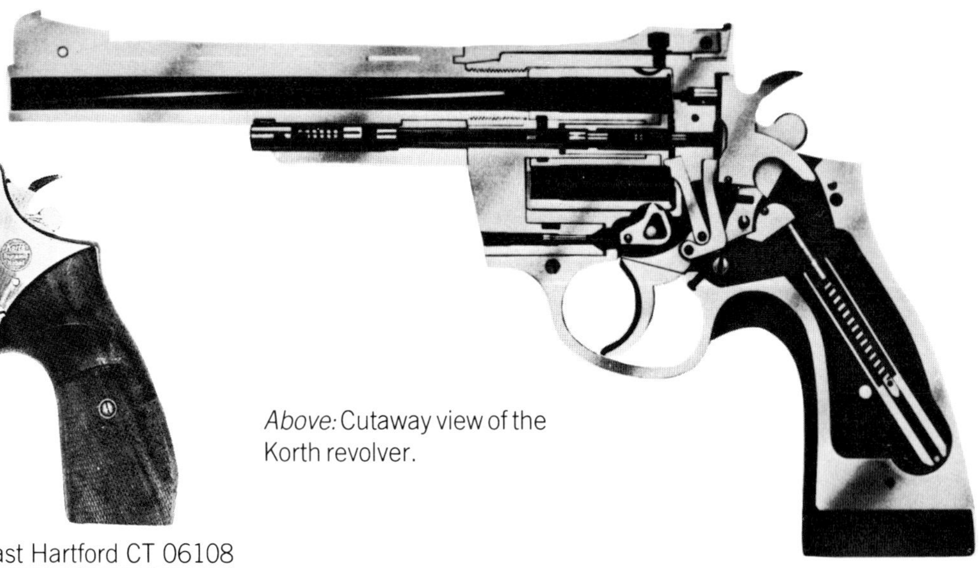

Above: Cutaway view of the Korth revolver.

HIGH STANDARD High Standard Inc., East Hartford CT 06108

NAME	CALIBER	OPERATION	LENGTH	WEIGHT EMPTY	BARREL LENGTH	MAGAZINE CAPACITY	REMARKS
Camp Gun	.22LR	Revolver	11.10in	28.0oz	5.30in	9	
Custom 10-X	.22LR	Blowback			6.75in	10	Target pistol

NAME	CALIBER	OPERATION	LENGTH	WEIGHT EMPTY	BARREL LENGTH	MAGAZINE CAPACITY	REMARKS
HIGH STANDARD Contd.							
Double Nine	.22LR	Revolver	11.00in	32.0oz	5.30in	9	
Sentinel	.22LR	Revolver	9.10in	27.0oz	5.30in	9	
Sharpshooter	.22LR	Blowback	10.30in	42.0oz	5.30in	10	Target pistol
Citation	.22LR	Blowback	11.60in	45.0oz	7.30in		Target pistol
Victor	.22LR	Blowback	9.60in	47.0oz	5.50in	10	Target pistol
IVER JOHNSON: Iver Johnson Arms Inc, Middlesex NJ 08846							
Pony X300	9mm/.380	Blowback	6.10in	20.0oz	3.00in	6	
KORTH: Waffenfabrik W. Korth, Ratzeburg-i-Holstein, West Germany							
Combat	.357M	Revolver		33.5oz	3.00in	6	
LLAMA: Gabilondo y Cia SA, Postbox 290, Vitoria, Spain							
Model IIIA	9mm/.380	Blowback	6.30in	23.0oz	3.70in	7	
Model VIII	.38Super	Recoil	8.50in	41.8oz	5.00in	9	Also 9mm Largo
Model IX-A	.45ACP	Recoil	8.50in	42.0oz	5.00in	8	
Model XXIII	.38SP1	Revolver	12.00in	40.7oz	5.50in	6	Target pistol
Comanche	.357M	Revolver	11.25in	35.6oz	6.00in	6	
Omni	9mmP	Recoil	7.90in	37.2oz	4.40in	9 or 13	Also .45 ACP

The Llama Comanche revolver, a well-made and accurate design.

NAME	CALIBER	OPERATION	LENGTH	WEIGHT EMPTY	BARREL LENGTH	MAGAZINE CAPACITY	REMARKS
MANURHIN: Manufacture de Machines du Haut-Rhin, 68060 Mulhouse, France							
MR73	.357M	Revolver	8.07in	31.4oz	3.00in	6	Also with other barrel lengths
MR32 Match	.38Spl	Revolver	10.90in	38.4oz	6.00in	6	Also .32
MAUSER: Mauser-Werke Oberndorf GmbH, D-7238 Oberndorf-am-Neckar, West Germany							
Revolver	.38Spl	Revolver	6.89in	21.2oz	2.50in	6	
RANDALL: Randall Firearms Co., Sun Valley, CA 91352							
Stainless .45	.45ACP	Recoil	8.50in	40.0oz	5.00in	7	
REMINGTON: Remington Arms Co., Bridgeport, CT 06602							
XP-100	.221Rem	Single shot	16.80in	60.0oz	10.50in	–	Also in 7mm caliber with 14.8in bbl
RENATO GAMBA: Armi Renato Gamba SpA, Postbox 48, I-25063 Gardone Val Trompia, Brescia, Italy							
Model 80 HSc	7.65mm	Blowback	6.53in	26.4oz	3.54in	15	Licensed Mauser design
Trident Super 4	.38Spl	Revolver	9.45in	25.4oz	3.97in	6	
Trident Fast Action	.38Spl	Revolver	6.85in	22.9oz	2.50in	6	

The Renato Gamba Trident .38 Special.

The Renato Gamba Model HSc Super is also available in 9mm Short and 9mm Police calibers.

The Manurhin MR73, available chambered for .357 Magnum and .38 Special ammunition.

NAME	CALIBER	OPERATION	LENGTH	WEIGHT EMPTY	BARREL LENGTH	MAGAZINE CAPACITY	REMARKS
ROHM: Rohm GmbH, D-7927 Sontheim-a-d-Brenz, West Germany							
RG-14	.22LR	Revolver	5.70in	14.8oz	1.80in	6	Also 3in bbl
RG-23	.22LR	Revolver	6.90in	15.2oz	3.00in	6	
RG-34	.22LR	Revolver	9.30	33.1oz	4.00in	6	
RG-34T	.22LR	Revolver	11.20in	43.0oz	6.00in	6	Target pistol
RG-57	.357M	Revolver	11.40in	40.2oz	6.00in	6	Also .41Mag, 44Sp .44Mag, .45Colt
RG-88	.357	Revolver	9.30in	32.3oz	4.00in	6	

NAME	CALIBER	OPERATION	LENGTH	WEIGHT EMPTY	BARREL LENGTH	MAGAZINE CAPACITY	REMARKS
ROSSI: Amadeo Rossi SA, São Leopoldo, Brazil							
Model 68	.38Spl	Revolver	8.50in	21.8oz	3.00in	5	
Model 94	.38Spl	Revolver	8.25in	25.0oz	3.00in	6	

Above: The Rossi Champion revolver.

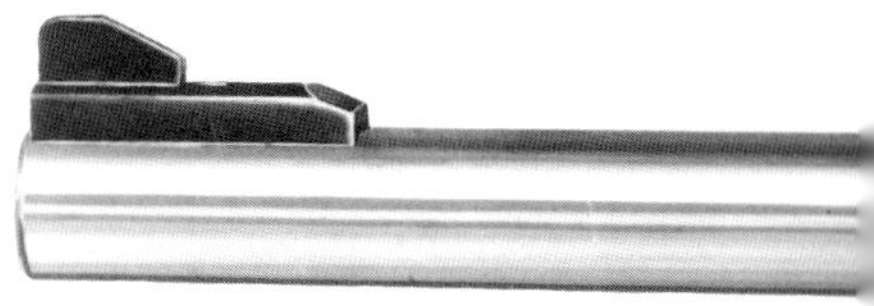

NAME	CALIBER	OPERATION	LENGTH	WEIGHT EMPTY	BARREL LENGTH	MAGAZINE CAPACITY	REMARKS
RUGER: Sturm Ruger & Co Inc., Southport, CT 06490							
Blackhawk	.357M	Revolver	12.30in	42.0oz	6.50in	6	Also .30 Carbine, .41Mag and .45 Colt
Super Blackhawk	.44Mag	Revolver	16.40in	51.0oz	10.5in	6	
Model GS-32N	.38Spl	Revolver	7.75in	34.0oz	2.75in	6	Also .357Mag and 9mm Parabellum

NAME	CALIBER	OPERATION	LENGTH	WEIGHT EMPTY	BARREL LENGTH	MAGAZINE CAPACITY	REMARKS
RUGER Contd.							
Police Service Six	.38Spl	Revolver	9.25in	32.9oz	4.0in	6	Also .357 Mag and 9mm Para
Redhawk	.44Mag	Revolver	13.00in	52.5oz	7.50in	6	
Security Six	.357M	Revolver	9.25in	33.0oz	4.00in	6	

Right: The Ruger Police Service Six.

Above: The Ruger Single-Six, shown here with 9.5-inch barrel.

NAME	CALIBER	OPERATION	LENGTH	WEIGHT EMPTY	BARREL LENGTH	MAGAZINE CAPACITY	REMARKS
RUGER Contd.							
Single Six	.22LR	Revolver	10.80in	33.0oz	5.50in	6	Also 4.6in, 6.5in, 9.5in barrels. Convertible to .22WMRF
Mark II	.22LR	Blowback	8.80in	36.0oz	4.80in	10	Also 6in bbl

The Ruger Mark II, with action open.

The Semmerling LM-4, a most unusual hand-operated repeater.

NAME	CALIBER	OPERATION	LENGTH	WEIGHT EMPTY	BARREL LENGTH	MAGAZINE CAPACITY	REMARKS
SAKO: Oy Sako AB, 11100 Riihimaki, Finland							
Convertible	.22LR	Blowback		44.1oz	5.90in	5	Match pistol
Convertible	.22 Short	Blowback		42.8oz	4.33in	5	Match pistol
Convertible	.32 S&W	Blowback		46.6oz	5.90in	5	Match pistol
SEMMERLING: Semmerling Corp., Newton, MA 02160							
LM-4	.45ACP	Blow-forward	5.10in	25.7oz	3.90in	4	

NAME	CALIBER	OPERATION	LENGTH	WEIGHT EMPTY	BARREL LENGTH	MAGAZINE CAPACITY	REMARKS
SIG-SAUER: Schweizerische Industrie-gesellschaft, CH-1212 Neuhausen-am-Rheinfalls, Switzerland, and J. P. Sauer & Sohn GmbH, Eckernforde, West Germany							
P-220	9mmP	Recoil	7.80in	28.5oz	4.41in	9	Also in .45ACP, .38 Super, 7.65mm Para and .22LR
P-225	9mmP	Recoil	7.09in	28.2oz	3.85in	8	Improved P-220
P-226	9mmP	Recoil	7.72in	29.6oz	4.41in	15	To US XM9 test requirements

Above: The SIG-Sauer P220 combat pistol has been adopted by the Swiss Army as the Pistole 75.

Above: The SIG-Sauer P226.

NAME	CALIBER	OPERATION	LENGTH	WEIGHT EMPTY	BARREL LENGTH	MAGAZINE CAPACITY	REMARKS
SIG-SAUER Contd.							
P-230	9mmPol	Blowback	6.61in	25.7oz	3.62in	7	Also in 9mm/.380 and 7.65mm

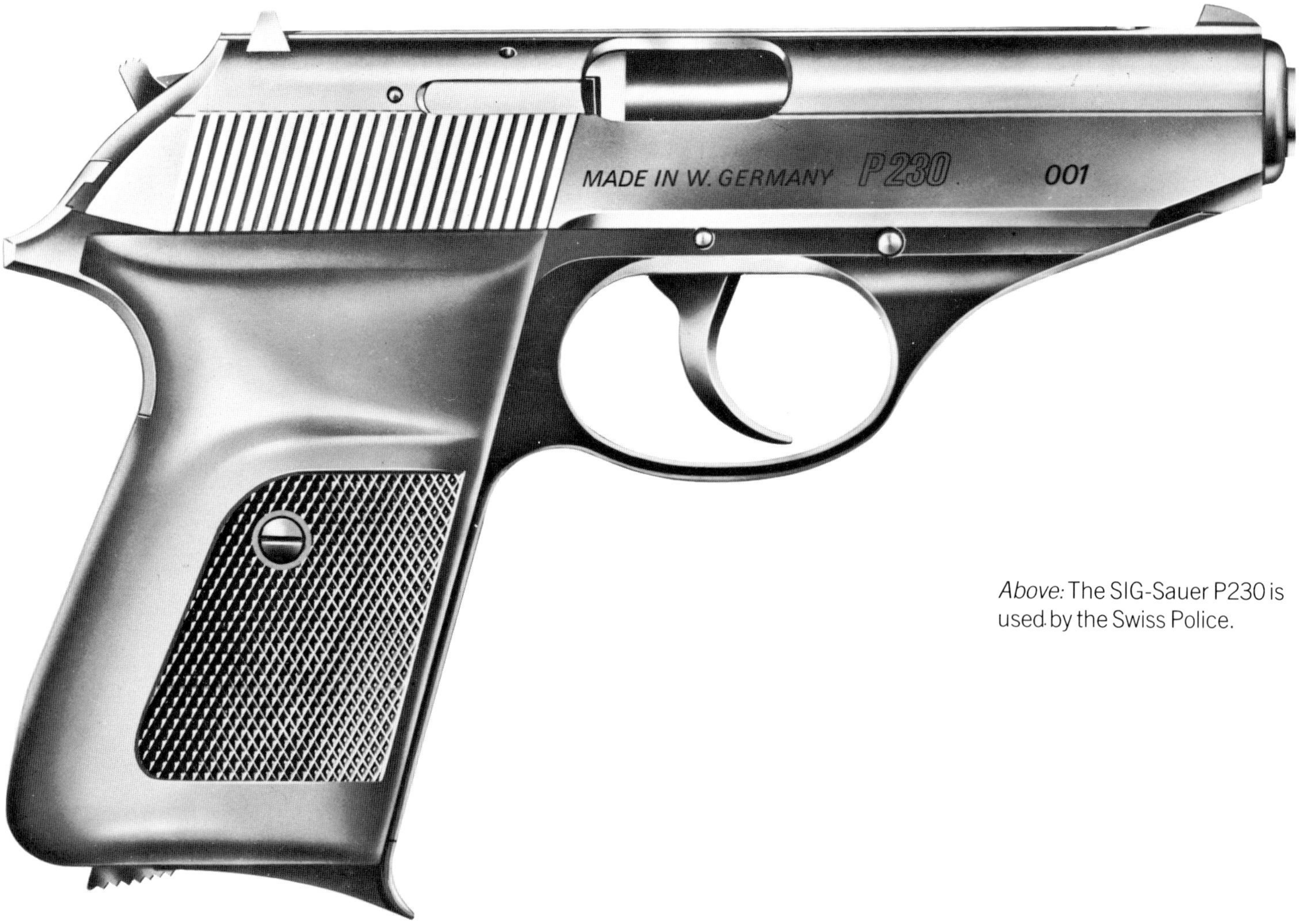

Above: The SIG-Sauer P230 is used by the Swiss Police.

SMITH & WESSON: Smith & Wesson, Springfield, MA 01101

NAME	CALIBER	OPERATION	LENGTH	WEIGHT EMPTY	BARREL LENGTH	MAGAZINE CAPACITY	REMARKS
Model 10	.38Spl	Revolver	11.30in	34.0oz	6.00in	6	
Model 19	.357M	Revolver	9.50in	35.0oz	4.00in	6	
Model 25	.45ACP	Revolver	11.40in	45.0oz	6.00in	6	Target pistol
Model 28	.357M	Revolver	11.30in	44.0oz	6.00in	6	
Model 31	.32S&W	Revolver	7.50in	22.0oz	3.00in	6	
Model 34	.22LR	Revolver	8.00in	24.5oz	4.00in	6	
Model 41	.22LR	Blowback	10.50in	42.0oz	7.00in	10	Target pistol
Model 48	.22LR	Revolver	11.10in	39.0oz	6.00in	6	
Model 52	.38	Recoil	8.60in	41.0oz	5.00in	5	Target pistol
Model 59	9mmP	Recoil	7.43in	27.0oz	4.00in	14	

The Smith & Wesson Model 586 .357 Distinguished Combat Magnum with 6-inch barrel.

The Smith & Wesson Model 17 K22 Masterpiece, a target shooting weapon in .22LR caliber.

NAME	CALIBER	OPERATION	LENGTH	WEIGHT EMPTY	BARREL LENGTH	MAGAZINE CAPACITY	REMARKS
SMITH & WESSON Contd.							
Model 459	9mmP	Recoil	7.43in	28.0oz	4.00in	14	Modified 59
Model 469	9mmP	Recoil	6.75in	42.0oz	3.50in	12	

The Smith & Wesson Model 681 .357 Distinguished Service Magnum with 4-inch barrel.

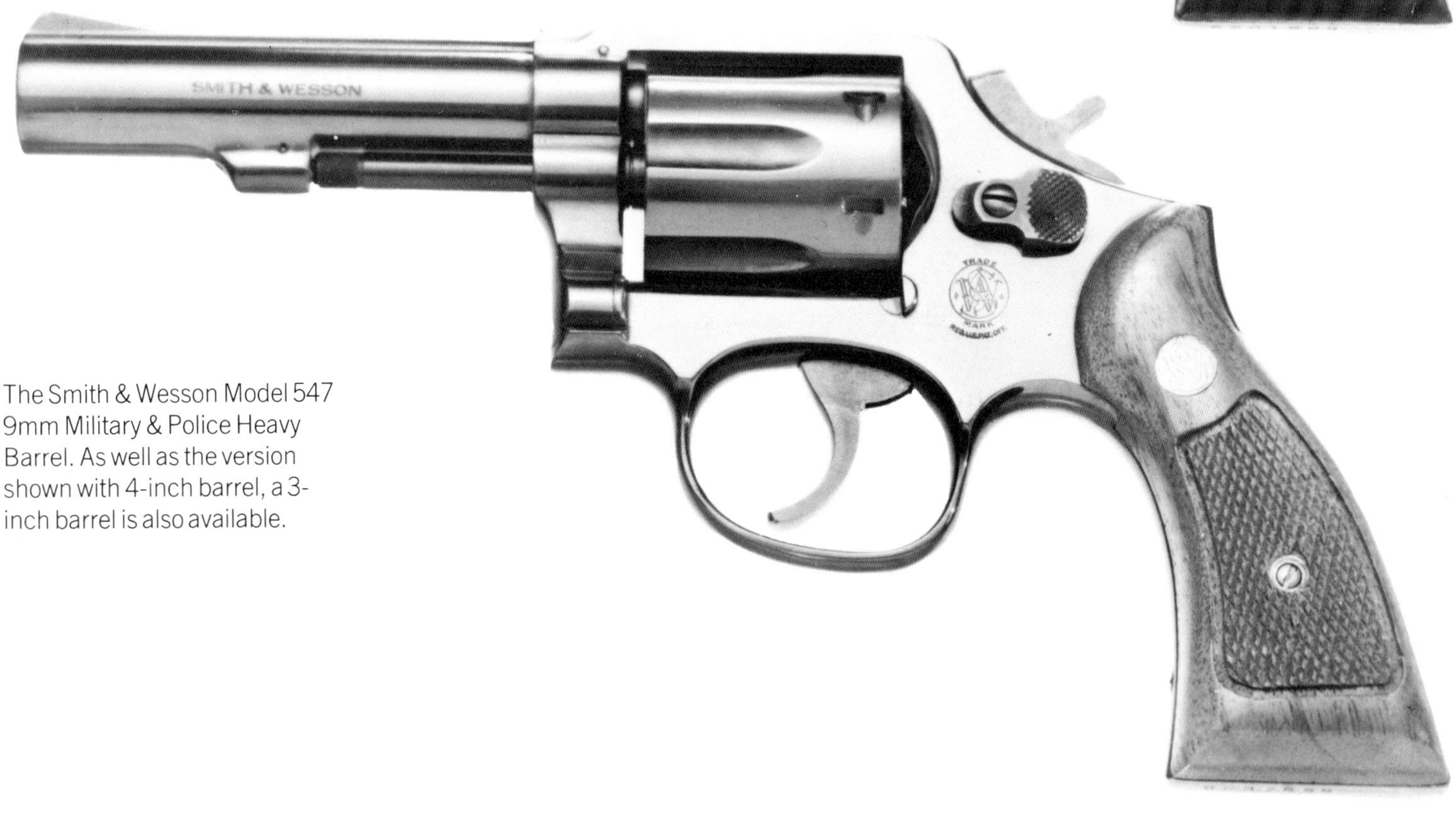

The Smith & Wesson Model 547 9mm Military & Police Heavy Barrel. As well as the version shown with 4-inch barrel, a 3-inch barrel is also available.

NAME	CALIBER	OPERATION	LENGTH	WEIGHT EMPTY	BARREL LENGTH	MAGAZINE CAPACITY	REMARKS
SQUIBMAN: Squires Bingham Mfg. Co., Manila, Luzon, PI							
Model 100DA	.38Spl	Revolver	8.30in	29.0oz	3.00in	6	Various barrel lengths

NAME	CALIBER	OPERATION	LENGTH	WEIGHT EMPTY	BARREL LENGTH	MAGAZINE CAPACITY	REMARKS
STAR: Star Bonifacio Echeverria SA, Postbox 10, Eibar, Guipuzcoa, Spain							
Super-B	9mmP	Recoil	8.74in	35.9oz	5.12in	8	
Model 28DA	9mmP	Recoil	8.07in	40.2oz	4.33in	15	Double action
Model DKL	9mm/.380	Recoil	5.70in	14.8oz	3.17in	6	Light alloy frame
Model BK	9mmP	Recoil	7.16in	34.0oz	3.89in	8	Also BKM with alloy frame
Model PD	.45ACP	Recoil	7.08in	25.0oz	3.94in	6	
Model FR	.22LR	Blowback	7.30in	26.0oz	4.30in	10	
Model SS	9mm/.380	Blowback	6.50	22.0oz	4.00in	8	

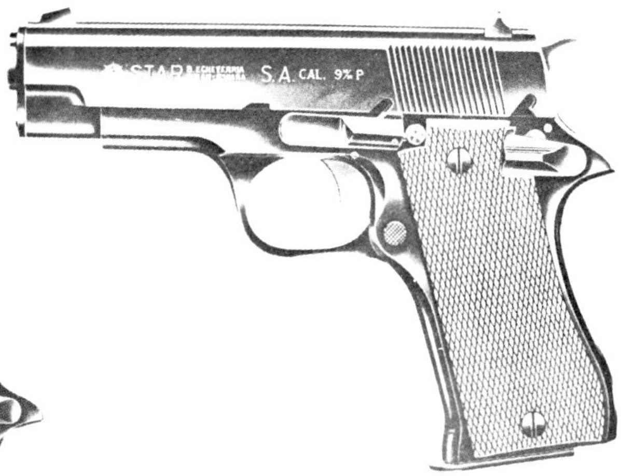

Above left: The compact and powerful Star Model BKM.
Left: The Star Model PD, one of the most compact .45 pistols available.
Below: The Star Model FR Target Pistol.

STERLING: Sterling Arms Corp., Lockport, NY 14094							
Model 300	6.35mm	Blowback	4.80in	13.0oz	2.00in	6	
Model 400	9mm/.380	Blowback	6.50in	26.0oz	3.75in	7	

NAME	CALIBER	OPERATION	LENGTH	WEIGHT EMPTY	BARREL LENGTH	MAGAZINE CAPACITY	REMARKS
STERLING Contd.							
Model 402	7.65mm	Blowback	6.50in	27.0oz	3.75in	7	
X-Caliber	.22LR	Single shot	13.00in	58.0oz	8.00in	–	Also 10in bbl; and in .22WMRF, .357Mag and .44Mag

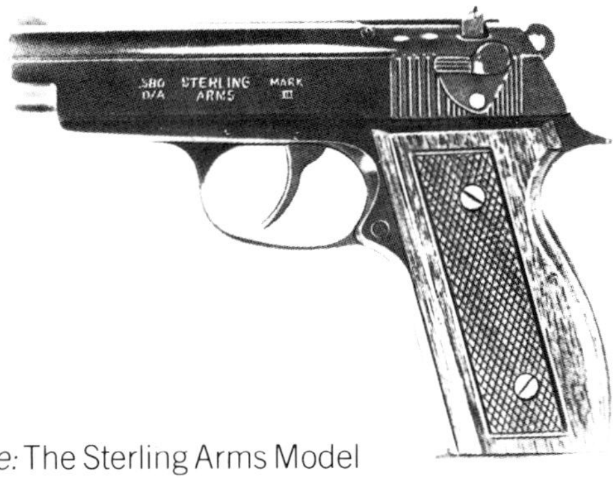

Above: The Sterling Arms Model 400 .380 automatic.

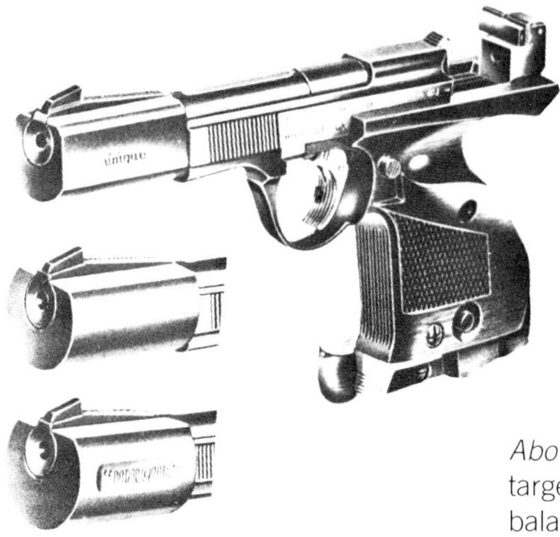

Above: The Unique DES-69 target pistol showing the various balance weights that may be fitted.

Above: The Steyr GB is made entirely of steel, even the grip plates.

Below: The Walther PP Super.

STEYR: Steyr-Daimler-Puch, Postbox 1000, A-4400 Steyr, Austria

NAME	CALIBER	OPERATION	LENGTH	WEIGHT EMPTY	BARREL LENGTH	MAGAZINE CAPACITY	REMARKS
Model GB	9mmP	Gas delayed	8.5in	29.8oz	5.35in	18	

NAME	CALIBER	OPERATION	LENGTH	WEIGHT EMPTY	BARREL LENGTH	MAGAZINE CAPACITY	REMARKS
THOMPSON: Auto-Ordnance Corp, West Hurley, NY 12491							
Thompson	.38Super	Recoil	8.50in	40.0oz	5.00in	9	Also .45ACP and 9mm Parabellum
THOMPSON/CENTER: Thompson/Center Arms, Rochester NH 03867							
Contender	Various	Single shot	13.50in	43.0oz	10.0in	–	Almost any desired caliber
U.S.S.R.: State Arsenals							
PSM	5.45mm	Blowback	6.29in	16.2oz	3.34in	8	Security police
TOZ-35	.22LR	Single shot	17.30in	45.0oz	11.80in	–	Free pistol
UBERTI: Aldo Uberti & Cia SpA, Ponte Sanano, Brescia, Italy							
Inspector	.32S&W	Revolver	8.00in	24.7oz	3.00in	6	
UNIQUE: Manufacture d'Armes des Pyrenees Francaises. F-64700 Hendaye, France							
DES-69	.22LR	Blowback	10.60in	35.3oz	5.90in	10	Target pistol
DES-VO	.22Short	Blowback	10.00in	43.7oz	5.90in	10	Target pistol
UZI: Israeli Military Industries, Postbox 1044, Ramat Ha Sharon, 47100 Israel							
Uzi Pistol	9mmP	Blowback	9.45in	66.0oz	4.53in	20	
WALTHER: Carl Walther Waffenfabrik, Postbox 4325, D-7900 Ulm, West Germany							
Model FP	.22LR	Single shot	17.30in	47.6oz	11.80in	–	Free pistol
Model GSP	.22LR	Blowback	11.50in	41.6oz	4.20in	5	Also .32S&W
Model P38K	9mmP	Recoil	6.30in	27.2oz	2.80in	8	Short P-38
Model P4	9mmP	Recoil	7.90in	28.4oz	4.30in	8	Short P-38
Model P1	9mmP	Recoil	8.50in	28.2oz	4.90in	8	Modern P-38
Model P5	9mmP	Recoil	7.08in	28.0oz	3.54in	8	
Model PP	7.65mm	Blowback	6.70in	22.6oz	3.90in	8	Also .22RF and 9mm/.380
Model PP Super	9mmPol	Blowback	6.90in	27.5oz	3.60in	7	Also 9mm/.380
Model TPH	.22LR	Blowback	5.30in	11.4oz	2.80in	6	
WILDEY: Wildey Firearms Co Inc., Cheshire, CT 06410							
WM450	.45WinM	Gas	11.00in	64.0oz	5.00in	8	Also 9mmWinM
ZASTAVA: Zavodi Crvena Zastava, Novembra 12, Beograd 29, Yugoslavia							
Model 70	7.65mm	Blowback	7.87in	26.1oz	3.70in	8	

Two comparative rarities nowadays; left, a shotgun/rifle combination and, right, a double rifle, both from Winchester.

RIFLES

Unlike pistols, which we have conveniently sub-divided into automatics, revolvers and others, rifles are still manufactured in a wide variety of types. While we can conveniently divide them into automatic, magazine and single shot groups, there are many sub-divisions within these, and even such an apparently obsolete pattern as the classic double rifle can still be obtained quite easily.

There have been a number of recent introductions in the military arena which suggest that after the flurry of activity in pistol design which has occupied the past ten years, there is now going to be something similar in the military rifle field. As is well known, after a long and tedious NATO trial a new 5.56mm cartridge, using a Belgian-designed bullet somewhat heavier and longer than the previous standard – the American M193 – was adopted, and already this is causing disquiet.

The Belgian SS109 bullet was selected because in the trials it proved to have better penetration and remaining velocity than the M193; in other words it would pierce a GI helmet (the readily understandable criterion of penetration) at greater ranges. Unfortunately, because it is longer and heavier than the M193 it demands a different twist of rifling in the weapon's barrel, which meant that any country adopting the bullet (and the NATO armies agreed to do just that) would have either to re-barrel their existing rifles or adopt some new rifle. As it happened, this was less of a hardship than it appears; most of the NATO armies were looking for a new rifle anyway, their 7.62mm weapons being twenty or more years old and approaching the time for scrapping. The people who would be hit most would be the US Army with a stock of over one million 5.56mm M16A1 rifles all of which had the wrong rifling twist.

But plans were already afoot for improving the M16A1 rifle, and these were extended to cover the adoption of the new twist of rifling, and the design has now gone into US service as the M16A2. In addition to the rifling change, there have been improvements in the sights, in the plastic fore-end and stock, the addition of a deflector to send ejected cartridge cases well clear of the face of a left-handed shooter, and the addition of some metal to the barrel to make it stronger. The M16A2 has cured some of the minor defects of its predecessor and can be expected to serve the US forces for the remainder of the century.

The complaints which survive appear to stem from incomplete understanding of the new ammunition. The most common observation is that the 5.56mm SS109 bullet is not the equal of the 7.62×51mm NATO standard bullet. Why anyone should expect it to be is far from clear, but the complaints still appear. There are also objections that the 5.56mm bullet is not particularly accurate at ranges over 400 yards, but, again, it is difficult to understand why anyone should expect it to be. The 5.56mm bullet was never intended to be a long-range projectile, nor is the M16 (or any other 5.56mm rifle) intended as a long-range sniping weapon. Provided the short-range application of the assault rifle is firmly grasped, then the rest of it falls into place. Where the long range penetration aspect of the SS109 bullet comes into play is in its use with 5.56mm machine guns, which is a totally different question.

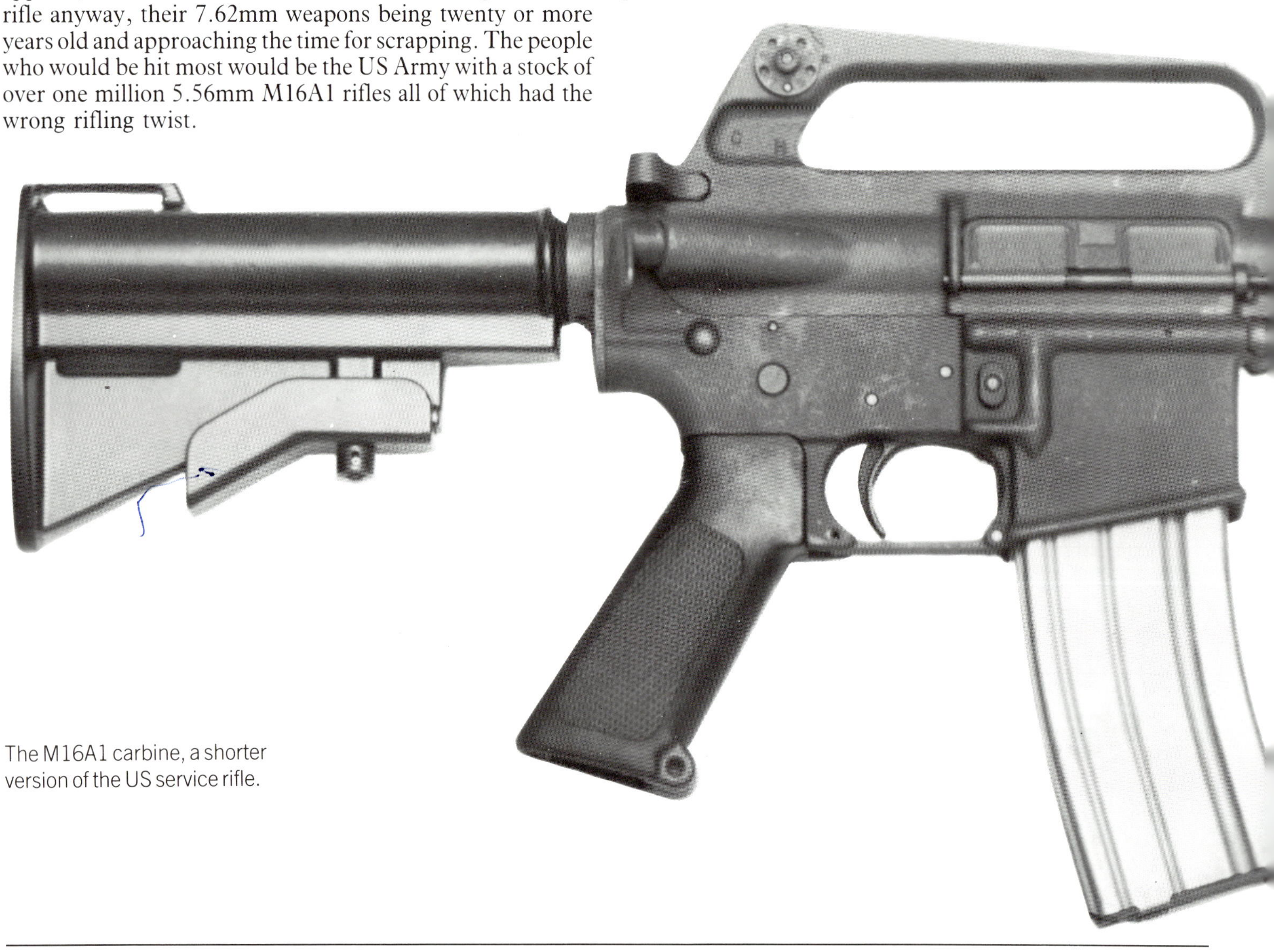

The M16A1 carbine, a shorter version of the US service rifle.

Another new feature on the M16A2 rifle is the incorporation of a ratchet device into the firing mechanism to allow a three-shot burst to be fired for one pull on the trigger. This has been adopted because the mathematics of hit probability show that there is an increased chance of hitting at maximum fighting ranges if three shots can be fired as rapidly as possible; they will be distributed slightly by the laws of probability and thus if the first shot misses there is a good chance that one of the next two will hit. However, mathematics do not take into account the behaviour of the rifle in the man's hands; experience shows that the recoil causes the weapon's muzzle to move upwards and to the right as the shots are fired, so that if the first shot hits the target dead center, then the next will hit high right and the third will miss, higher and more right. In an endeavor to avoid this the M16A2 has had the muzzle compensator and flash eliminator modified by omitting the holes which pointed downwards. This helps to counter the lift, since there is no downward escape of gas to push the muzzle up, in fact there is now only an upward escape which will help to keep the muzzle down. Whether this has as good an effect as some people aver is yet to be made clear; there is more to the lift of the muzzle than gas escape. One of the reasons for the muzzle movement is the lever action of the recoil forcing the butt against the firer's shoulder, causing the muzzle to rise, and the escaping gas is not going to prevent this. Another factor is the torque effect of the bullet resisting the rifling.

Opinions over the utility of the three-round burst mechanism – as a general concept, not as it relates to the M16A2 - are mixed. It is not without interest to note that the Spanish development agency CETME (Centro de Estudios Tecnicos de Materiales Especiales) developed a new 5.56mm rifle for the Spanish Army with a three-round burst facility, and then withdrew the rifle and re-designed it without the burst mechanism. When asked why, they replied that it was an additional piece of machinery to go wrong, and that a properly trained soldier could trigger a three-round burst quite comfortably without needing any special device to do it for him. It may be significant that the CETME Model L fires at a maximum rate of 800 rpm on automatic, while the M16A2 has a top rate of 940 rpm, which suggests that it might be easier for skilled operators to get off a consistent three-round burst with the slower rate of fire.

The only rifle to achieve the mathematically-perfect three-round burst is the Heckler & Koch G-11, currently under development for adoption by the West German Army in 1990. When the Bundeswehr laid down their requirements for their next-generation rifle, one of them was that it should fire a three-round burst and that the three bullets should have a maximum dispersion of 1.5 to 2.0 mils, depending upon the range. For those unfamiliar with the mil system of angular measurement, one mil is the angle made by a one-meter length at 1000 meters range - or one mile at 1000 miles, or one anything at 1000 anything. So in effect the Bundeswehr were

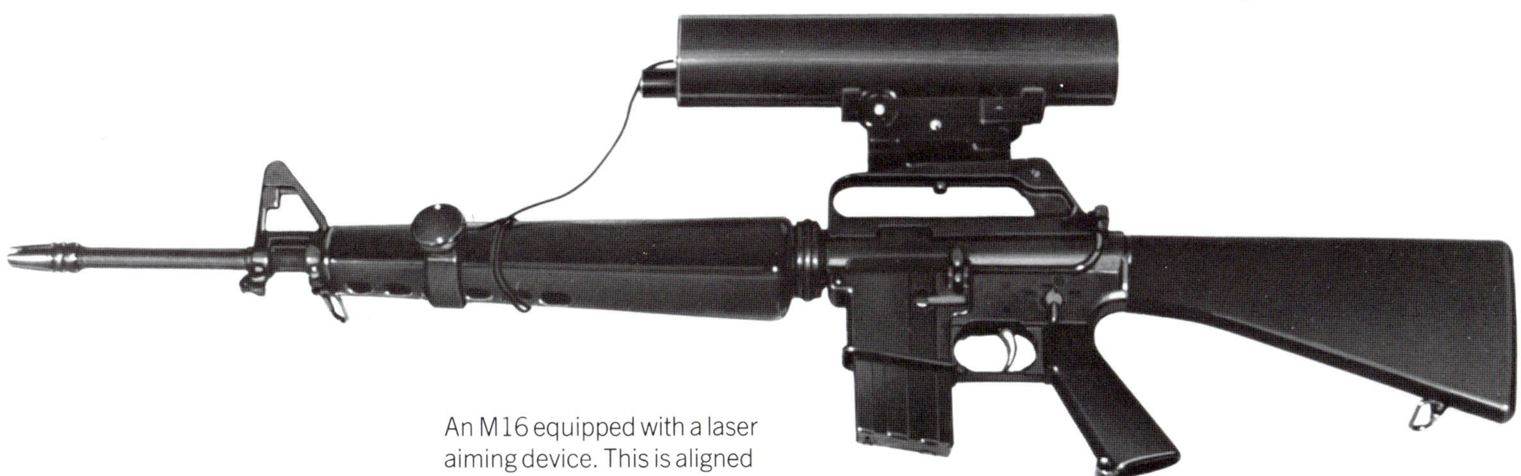

An M16 equipped with a laser aiming device. This is aligned with the bore so that the firer can detect the exact aiming point.

Above: Cutaway illustration of the mechanism of the Ruger Mini 14.
Right: The new British service rifle, the 5.56mm Individual Weapon with the standard 'Trilux' sight and accessories.

Heckler & Koch's revolutionary G11 rifle.

asking for three bullets to strike within a six-foot circle at 1000 yards range or a three-foot circle at 500 yards. It didn't take Heckler & Koch long to determine that there was no way in which a conventional weapon would meet this specification; the muzzle rise between shots made sure of that. Moreover mathematical studies soon showed that to achieve the desired 2.0 mil dispersion it would be necessary to fire the weapon at a rate of about 2000 rpm, and no known weapon system would do that with any accuracy. So they had to set about designing a totally new mechanism.

Exactly how their mechanism works is not publicly revealed, but they have given some information on the basic principles involved. Firstly they adopted a caseless cartridge, so as to do away with the problem of extracting the empty case between shots and thus save a considerable amount of time. Next they devised a rotating chamber at the breech end of the barrel; this when set with the chamber vertical, accepts a cartridge from the magazine (mounted above the barrel and feeding downwards). It then rotates a quarter-turn to line up the cartridge with the barrel, and the cartridge is then fired by a conventional firing pin. The barrel, breech and magazine, as a unit, then begins to recoil inside the casing of the rifle. As it does so the breech block rotates back, takes a new round, rotates again and fires it; this adds some momentum to the recoiling mass. The breech then makes a third rotation, loads, aligns, and fires the third shot, adding more momentum to the recoil. The recoiling parts are then allowed to complete their rearward stroke, come to rest on a buffer, and return to the initial forward position. What it means in practice is that three shots are fired for one pressure of the trigger and for one recoil movement of the barrel, and the three shots are out of the muzzle before the recoil stroke is completed and therefore before the force is transmitted to the firer's shoulder and before the gun muzzle can begin to rise.

The three shots sound almost like one, and the three bullets slap through the target separated by a few inches at most, well within the 1.5 mil dispersion demand. The actual rate of fire for those three shots is about 2200 rpm and they are long gone by the time the firer feels any recoil.

As might be imagined, the amount of research which has gone into this rifle is enormous; Heckler & Koch began working on it in about 1969; their first task was to develop a workable caseless cartridge, in conjunction with Dynamit Nobel, and in 1972 they began work on the rifle. By 1977

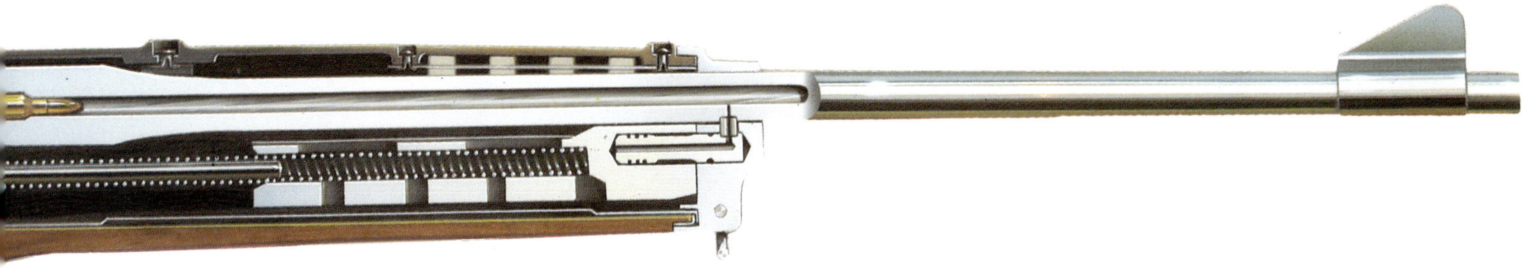

they had Prototype No 4 under test, by 1982 they had made about 60 prototypes, and by 1986 they hope to have the weapon production-engineered so that troop testing can get under way. The 1977 model had 341 parts and weighed 12.5lbs; by 1982 they had this down to 155 parts and 9.4lbs, and their aim by 1986 is to get it down to 100 parts and 7.9lbs.

The technical success of the G-11 rifle has raised interest in the USA and Heckler & Koch are currently working with the US Army on an 'Advanced Combat Rifle' program, using caseless technology. One prototype rifle from this program, seen recently, resembles the G-11 in external form – a plastic casing containing the weapon – and is chambered for an experimental 4.95mm caseless cartridge about which nothing has been disclosed. At present, though, several possible solutions are being explored by the US Army via various contractors, and it is unlikely that anything positive will be seen before 1987 or so.

One of the remarkable things about current military rifle development is the number of designs which are adopting the 'bull-pup' configuration. This puts the breech at the rear end of the weapon, close beneath the firer's cheek, and positions the pistol grip and trigger ahead of the magazine. The advantage is that a rifle built in this manner can have a full-length barrel in the shortest overall length of the weapon, so giving the advantages of compactness without sacrificing the accuracy of the weapon. The idea was pioneered in the late 1940s by the British, with their EM2 rifle, but this was not adopted for service; there was nothing wrong with the rifle, but the cartridge around which it had been designed failed to gain NATO acceptance, and the rifle could not be re-engineered to suit the NATO 7.62mm cartridge. However, when the time came to design a new rifle for the NATO trials in 1977 the British again put up a bull-pup, the Enfield XM-70. Again they had designed it around a totally new cartridge in 4.85mm caliber, and again NATO refused it, deciding on the 5.56mm round instead. But the astute British had developed their 4.85mm cartridge around the necked-down 5.56mm case, so that it was comparatively easy to re-design the XM-70 to fit the 5.56mm round. As a result the new British rifle, which is going into service in 1985, is a bull-pup.

The French have also got one in service, the FA-MAS, also in 5.56mm caliber. This actually went into troop use before the British design, and has seen active service in Lebanon; and, as usual, the French were not averse to selling it abroad

Right: The French Army's FA-MAS 5.56mm assault rifle.

Left: A French soldier using the FA-MAS rifle, fitted with an infra-red aiming spotlight and wearing infra-red goggles to permit accurate aiming in darkness.

Below: The Israeli Galil 5.56mm rifle with a range of its accessories.

before their own army was equipped, so that at least two armies in Africa have the weapon in use. The FA-MAS is an unusual-looking weapon; with a very large carrying handle and sight on top, which gives rise to the French soldiers' nickname for it – the Bugle. It is also unusual in being a delayed blowback design; a delay lever is interposed between bolt and bolt carrier, and this has to be rotated, against a mechanical disadvantage, to move the bolt carrier back and rotate the bolt, so giving a short delay which is enough to permit the bullet to leave the barrel.

The next bull-pup design to appear was from Finland, where the Valmet company adapted their standard M76 rifle mechanism into the new shape. The M76 is, in fact, little more than a somewhat improved Soviet Kalashnikov AK-47 rifle, the changes being little more than better plastic furniture, an aperture rear sight and a better flash hider. Revamping it into bull-pup form has meant raising the sights and re-designing the magazine, but the essential gas-operated mechanism remains the same. Whether the Soviets will look on this idea with any interest remains to be seen; Valmet claim that the rifle is in production, but there is no information forthcoming about who is buying it.

It looks as if the next bull-pup is going to come from Brazil. Brazil has been developing its armaments industry apace during the past decade and a new assault rifle is one of the principal targets. The 'Lapa 03' is expected to go into production in 1985, though this may prove to be an optimistic estimate. The mechanism is enclosed in a tough one-piece plastic body which runs from the butt to the fore-end and has a carrying handle and sight unit on top. The rifle is gas-operated in the conventional way, a gas piston above the barrel propelling a bolt carrier which rotates the bolt by a cam. An unusual feature is a double-action hammer mechanism; turning the selector switch to 'DA' drops the hammer safely and the rifle can then be carried with a round in the chamber. When required, a long pull on the trigger will raise and drop the hammer to fire the chambered round, after which the operation is normal. This system means that no safety catch is required. There is also to be a three-round burst facility as well as full-automatic fire or single shots.

The other interesting area of military rifle development is in sniping rifles. The standard sniping rifle, for many years, has usually been a bolt-action weapon based on the army's last manual rifle; thus, the British used their Lee-Enfield chambered for 7.62mm NATO, the US used Springfield M1903s for years and also a number of semi-automatic Garands, the Germans used a Mauser, the Austrians a Mannlicher and so on. This was due to the conviction that bolt-action rifles were more accurate than semi-automatics, and there is no doubt that this was a perfectly correct assessment if you were talking about the run-of-the-mill military semi-auto. Military accuracy standards, for service rifles, are not

Main picture: The Valmet Model 255-470, a 7.62mm bullpup design.
Top, far right: The Soviet 5.45mm family; top, the AK-74 rifle and, below, the RPK-74 light machine gun.

Bottom right: The shorter-barrel version of Fabrique Nationale's FNC 5.56mm rifle.

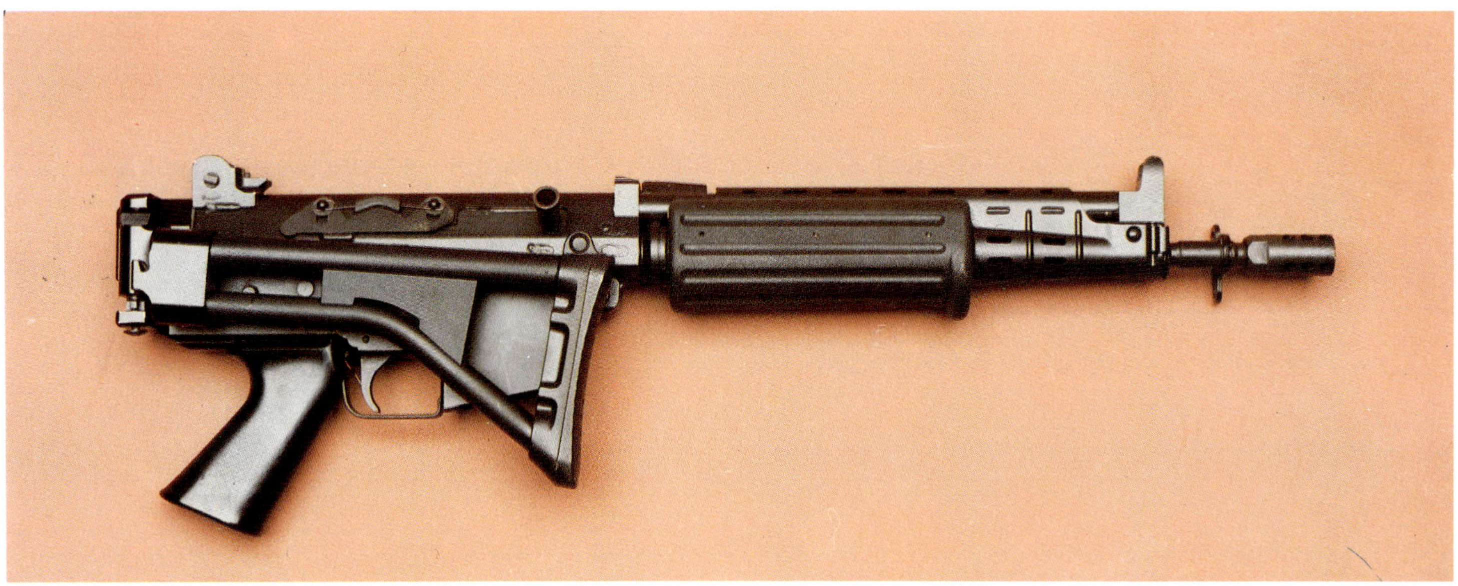

up to sniping standards, due to the clearances and tolerances incorporated so that the rifle will continue to work in every conceivable condition of climate and hard use. The old bolt-action rifles, made in traditional 'gunsmith' ways, with machined receivers and carefully-fitted bolts, were to tighter tolerances and by selecting the best ones at proof testing stage they could be worked over to put them into the sniping class. But the present-day stamped metal and plastic assault rifle is incapable of being re-worked except at considerable expense; the only people who have successfully modified their service rifle into a sniper are the Soviets with their Dragunov, which is a Kalashnikov action re-worked to accept the 7.62mm M1891 rimmed cartridge – and even then, according to Western experts who are familiar with the Dragunov, the improvements are marginal.

Today, however, the tendency is not to argue about bolt-action versus semi-auto, but simply to settle for a rifle which is built for the job of sniping – not just a converted or 'accurized' military model. A good example of this approach is the Mauser SP66, adopted by the Bundeswehr and at least 12 other armies. This is a bolt-action magazine rifle, but any similarity with a military Mauser rifle ends right there. To begin with, the bolt is the 'short-action' Mauser bolt, with the handle at the forward end, just behind the locking lugs; by adopting this type of bolt the movement is reduced by about 3½ inches, allowing this much to be added to the barrel. Another advantage is that when the bolt is opened there is less overhang and the firer doesn't have to move his head out of the way.

The barrel is heavier than normal, giving stiffness for accuracy, and is fitted with a muzzle brake and flash hider of considerable efficiency. The muzzle flash not only gives away the sniper's position, it also tends to blind him and in these days of image-intensifying sights or high-magnification telescopes, this can have serious effects on successive shots. The stock is a 'thumb-hole' patttern, and is fully adjustable for length and height of cheek-piece, and every surface to be touched by the hands is roughened to give a secure grip.

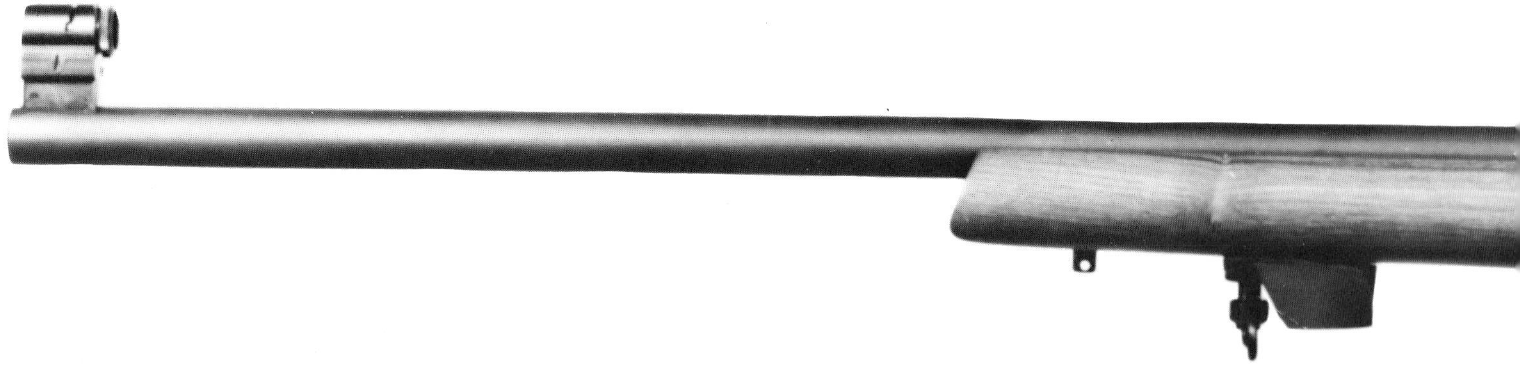

Above: The Galil 7.62mm sniper's rifle.
Right, main picture: The Parker Hale Model 82 7.62mm sniping rifle.

The Heckler & Koch PSG-1 7.62mm sniping rifle.

There are no iron sights, the issue sight being a Zeiss 1.5 to 6× zoom telescope. Altogether the SP66 is a weapon which would not look out of place in an international competition.

Another Mauser-based sniper is the Fabrique National Model 30-11, which was developed with an eye on police and security requirements rather than on army needs. This again has a heavy barrel with a good flash hider, and a stock which can be adjusted by means of inserts so as to fit any firer. It also has a bipod permanently mounted under the fore-end, which is particularly useful when heavy night vision sights are being used.

However, the semi-automatic, made for the task of sniping rather than simply adapted from a service pattern, is now beginning to appear. The first significant model was the Galil Sniper developed in Israel. The 5.56mm Galil rifle is well-known; what is less well-known is that it was then re-worked into 7.62mm caliber for those who preferred the heavier cartridge or for those armies which already had large stocks of 7.62mm and required a new rifle. This basic mechanism was then taken as the foundation for a specially-designed weapon solely for sniping. The barrel is much heavier than normal, and has a muzzle brake and flash hider; this can be removed and a silencer fitted if required. There is a bipod, attached to the receiver so that no strain is placed on the barrel and it is close enough to the pistol grip to permit the firer to adjust it easily. There is a massive cast sight-supporting bracket mounted on the side of the receiver to take a telescope or night sight, and this has been engineered so that the sight can be removed and replaced without affecting the zero of the rifle. The mechanism permits only single shots, and there is a two-stage trigger which allows lost motion to be taken up and then a crisp let-off which is absolutely regular and can be adjusted to the user's choice. The butt folds alongside the receiver, though this is intended simply to reduce the bulk for transport and not as a means of turning the weapon into a 'compact' design. Finally the butt-stock is fully adjustable and has a cheek-piece and a rubber recoil pad.

The most advanced sniping rifle to appear so far is the German Walther WA2000, a most remarkable weapon by any standards. In the first place the makers have ignored military standards in selecting the cartridge upon which the rifle is based; they have adopted the .300 Winchester Magnum, since their tests have shown that this is the most consistent in performance, giving the minimum round-to-round variation. They admit that this might cause logistic problems, however, and can produce the rifle in 7.62 NATO or 7.5mm Swiss Army chambering if so desired.

The construction is also unique, and owes nothing to any

The Ruger Model 77 Ultra-light bolt-action weapon.

Ruger's .223 Mini-14 uses the Garand system of gas operation and rotating bolt and is both light and reliable.

previous rifle, military or civilian. Since the barrel is the most important component, the whole rifle is built around a profiled barrel set into a rigid frame consisting of upper and lower arms. The muzzle end of the barrel is supported by two connected brackets between the upper and lower frame arms, and the arms are linked by side-plates at the rear end which form the receiver. This method of assembly creates a supportive structure in which the barrel is accurately located and in which the recoil force is delivered to the firer's shoulder in a perfectly straight line.

The barrel is carefully dimensioned and rifled so as to suit the particular bullet to be used, and it is externally fluted longitudinally, making it resistant to vibration and harmonic effects as well as increasing the surface area as an aid to cooling. It can be removed from the frame quite easily and replaced by one of a different caliber, though in such changes the bolt would also have to be changed. The bolt locks by seven lugs, rotated by a bolt carrier driven by a gas piston carried in one of the frame arms which doubles as the gas cylinder. The weapon is hammer fired, and the entire trigger

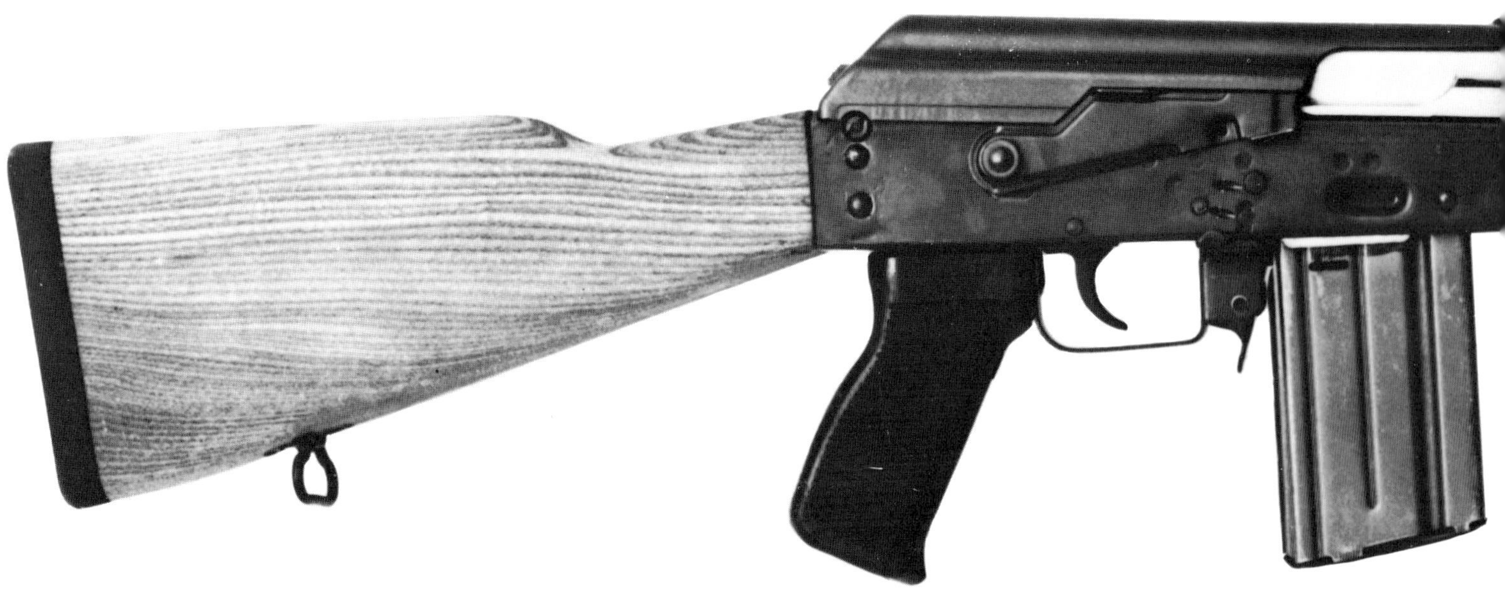

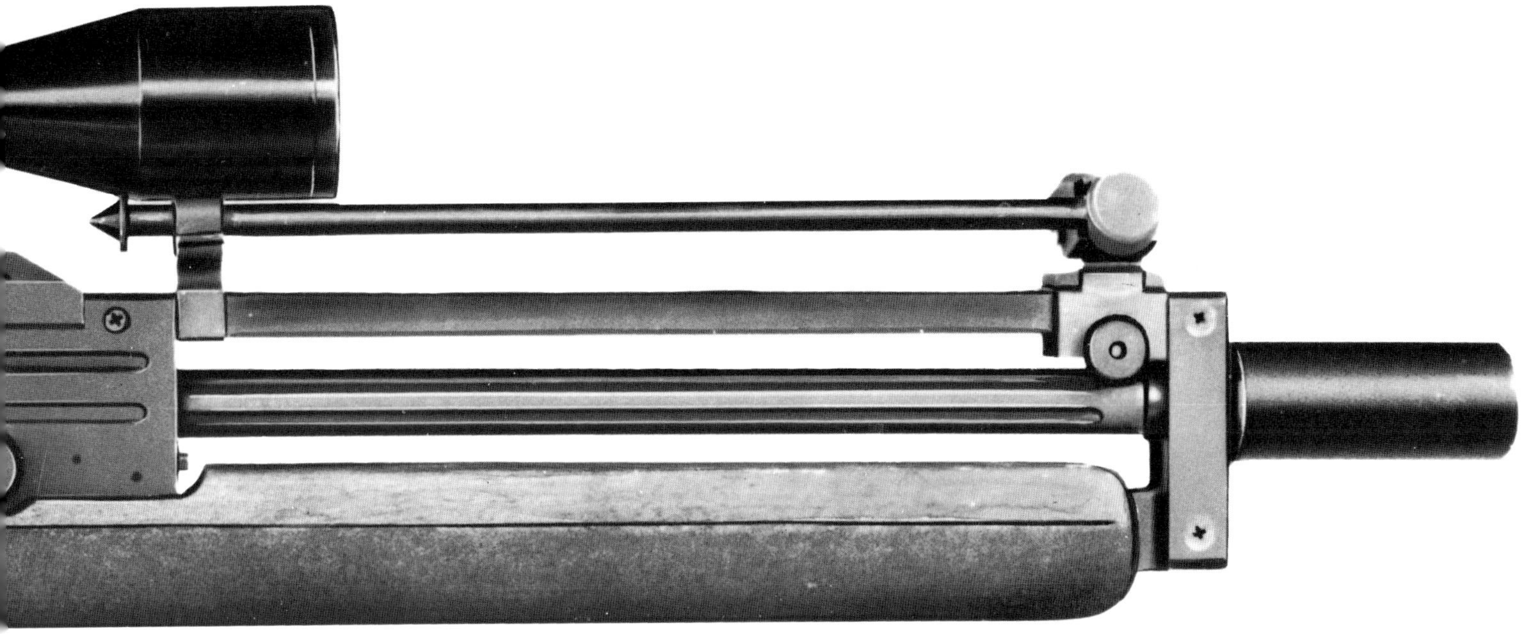

The highly developed Walther WA2000 sniping rifle.

and hammer unit can be slipped from the frame for lubrication, checking and adjustment. The standard sight is a Schmidt & Bender 2.5 to 10× zoom telescope, though any other sight can be fitted to the universal mounts. The WA2000 is available on the commercial market for those who can afford the necessary $3500 or so.

An odd aspect of sniping has made its appearance recently in the USA, the use of the .50 Browning machine gun cartridge as a sniping round. The prospect of using the .50 has often appealed to experimenters; a British officer in Korea, in 1952, had a spare .50 barrel stocked and breeched with a Mauser bolt action by a Japanese gunsmith, and used it for occasional sniping forays. He was intensely unpopular with the forward troops since after he had enjoyed a few shots and departed, the victims would proceed to retaliate with mortar fire and similar unpleasantness, which was rough on the residents.

At least three .50 sniping rifles are currently being offered in the USA, two being bolt action and one a semi-automatic using a recoiling barrel to provide the motive power. The

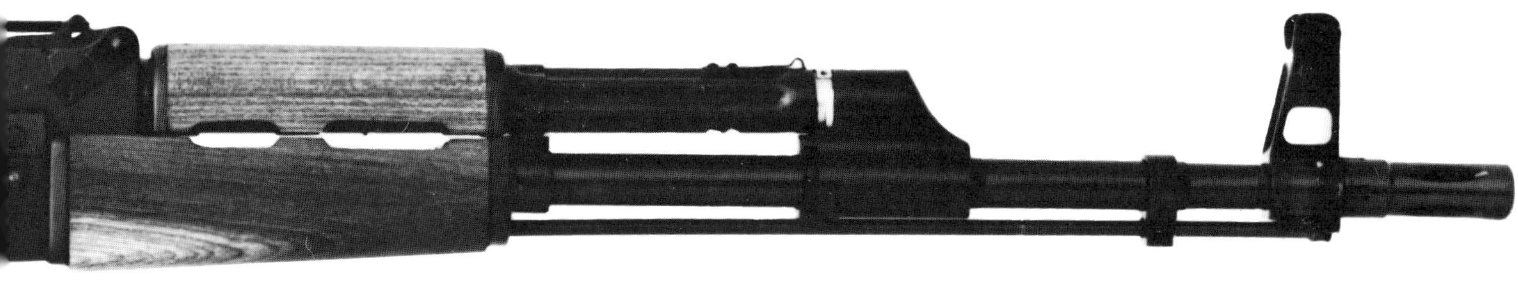

The Yugoslavian Zastava M80 5.56mm assault rifle.

Above: The Sterling Model 81 sniping rifle in 7.62mm caliber is made with a minimum of frills using the well-tried Mauser-type bolt and a very heavy barrel.

lightest of the three tips the scale at 26lbs, the heaviest at 35lbs, so none of them is exactly convenient to carry. But the drawback to this idea seems to be the ammunition problem; the 710 grain .50 Browning ball bullet does not have the degree of accuracy thought necessary for sniping; British acceptance specifications call for an 'Accuracy Figure of Merit' of 13 inches at 500 yards range. The 'Figure of Merit' is a peculiarly British method of assessment, and is the average distance in inches of twenty shots from the mean point of impact of the group, so an average of 13 inches at 500 yards is hardly pin-point accuracy.

Turning now to the sporting scene we can continue with our previous theme by starting with a look at the overlap between the military and civil markets, those rifles which began life with military intent but which, due to demand, have been released to the civil market. Although, as we have said, military specifications are drawn up to cover the use of the rifle in the worst possible conditions and not to demand extremes of accuracy, by careful selection of components and perhaps some special manufacture or finishing work here and there, it is possible to produce a weapon sufficiently accurate for hunting use without departing too far from the military spec. The principal change which has to be made, and which frequently involves quite a bit of expense, is changing the mechanism so that the rifle will only fire semi-automatic and, what is perhaps more important, cannot be 'home gunsmith' converted back to automatic action. More than one manufacturer has looked at the possible commercial market, costed the modifications necessary to meet legal requirements, and then turned the idea down since the cost of making the necessary changes would price the weapon out of what is, in fact, a fairly expensive market. So if you cannot find your favorite assault rifle advertised in semi-auto version, think about that for a moment.

The demand for military rifles is understandable; their robustness is beyond argument, since before their adoption they were put through tests which would make most commercial rifles lie down and die. For hunters and survivalists, therefore, they have a practical attraction. There is also the

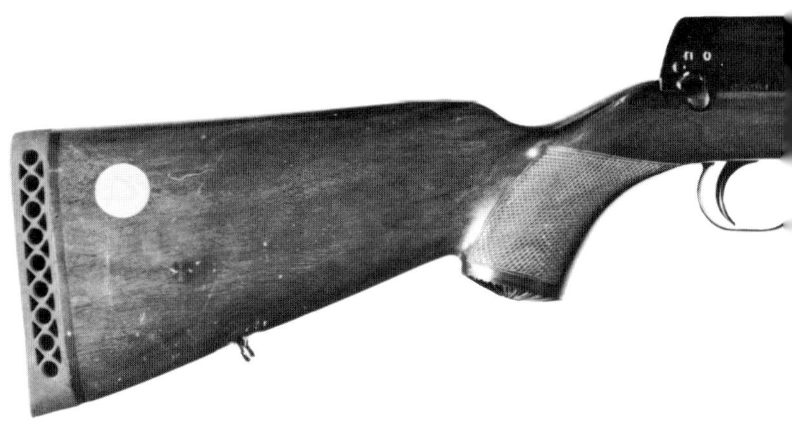

The Colt AR15 A2 Sporter II, a 'military style' rifle adapted for sporting use.

glamor of the military-style weapon with its olive drab plastic furniture, matt black or phosphated metal, big magazine, flash hider, grenade launcher, bayonet boss and so forth, a totally different attraction to the sleek lines of the traditional sporting rifle, all walnut and blue steel. But the accuracy aspect must be constantly borne in mind by those considering a military over a commercial weapon. Consider what the military rifle is meant to do, and then consider what you intend asking it to do; the soldier hunts one game and one game only, and his rifle is optimized for that purpose.

The most attractive things are, of course, those which are difficult to come by, and therefore the recent announcement by a Chinese company is almost certain to attract customers. The 'China North Industries Corporation' (Norinco) of Beijing (which is actually the commercial sales department of the Chinese Government arsenals organization) has begun offering both the Kalashnikov and Simonov semi-automatic rifles as sporting weapons. The Kalashnikov copy is called the 'Type 56S' and has been modified to fire semi-auto only and, presumably, so altered that it cannot be changed back. Instead of the usual laminated plywood furniture the fore-end, pistol grip and stock are of well-finished walnut, and the metal is blued to a good standard. As well as the normal fixed-butt version there are two folding -butt models, one with the steel strut stock folding downwards and forwards in the familiar AK47 manner, the other with a skeleton butt, partially filled with wood, folding sideways to lie alongside the receiver.

The Simonov copy is called the Type 56 (which could be a source of confusion) and is available in two versions. Both are mechanically the same, differing only in accessories and finish. The first is the standard Simonov with a 10-shot magazine, finger grooves in the fore-end and ventilation slots in the handguard, and a checkered walnut pistol-grip stock. The other has a five-shot magazine, a pistol grip stock of lighter wood – possibly sycamore – with more checkering at the fore-end instead of the finger grooves, no ventilation in the handguard, no sling swivels and a recoil pad. Unfortunately the Norinco literature makes no distinction between the two models, so anyone seeking a Simonov should be careful what he specifies.

Both Norinco weapons – the 56 and 56S – are chambered for the 7.62×39mm M1943 Soviet cartridge, which is not yet manufactured by any US company but which will doubtless

The Russian 'Medved' sporting rifle.

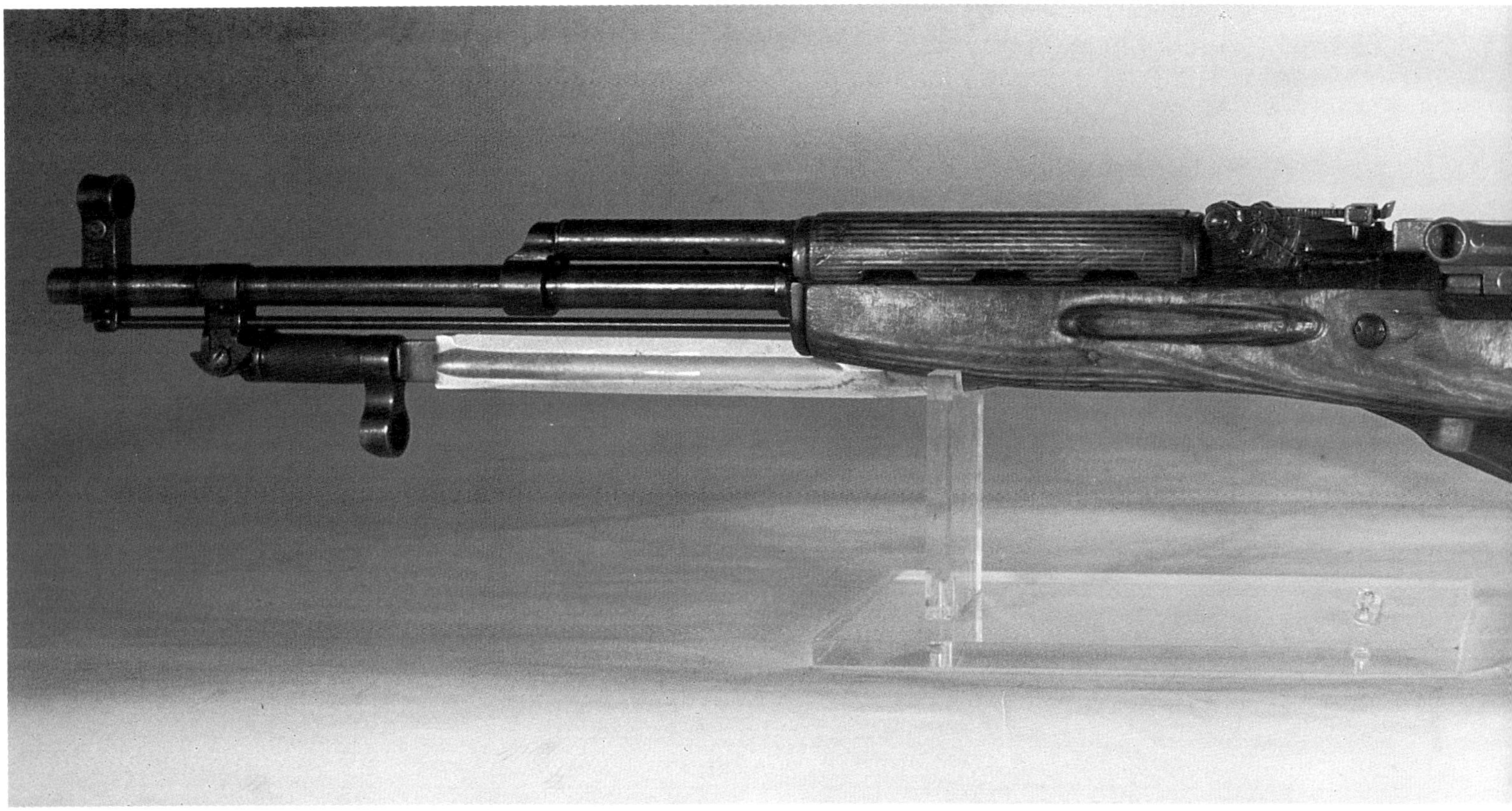

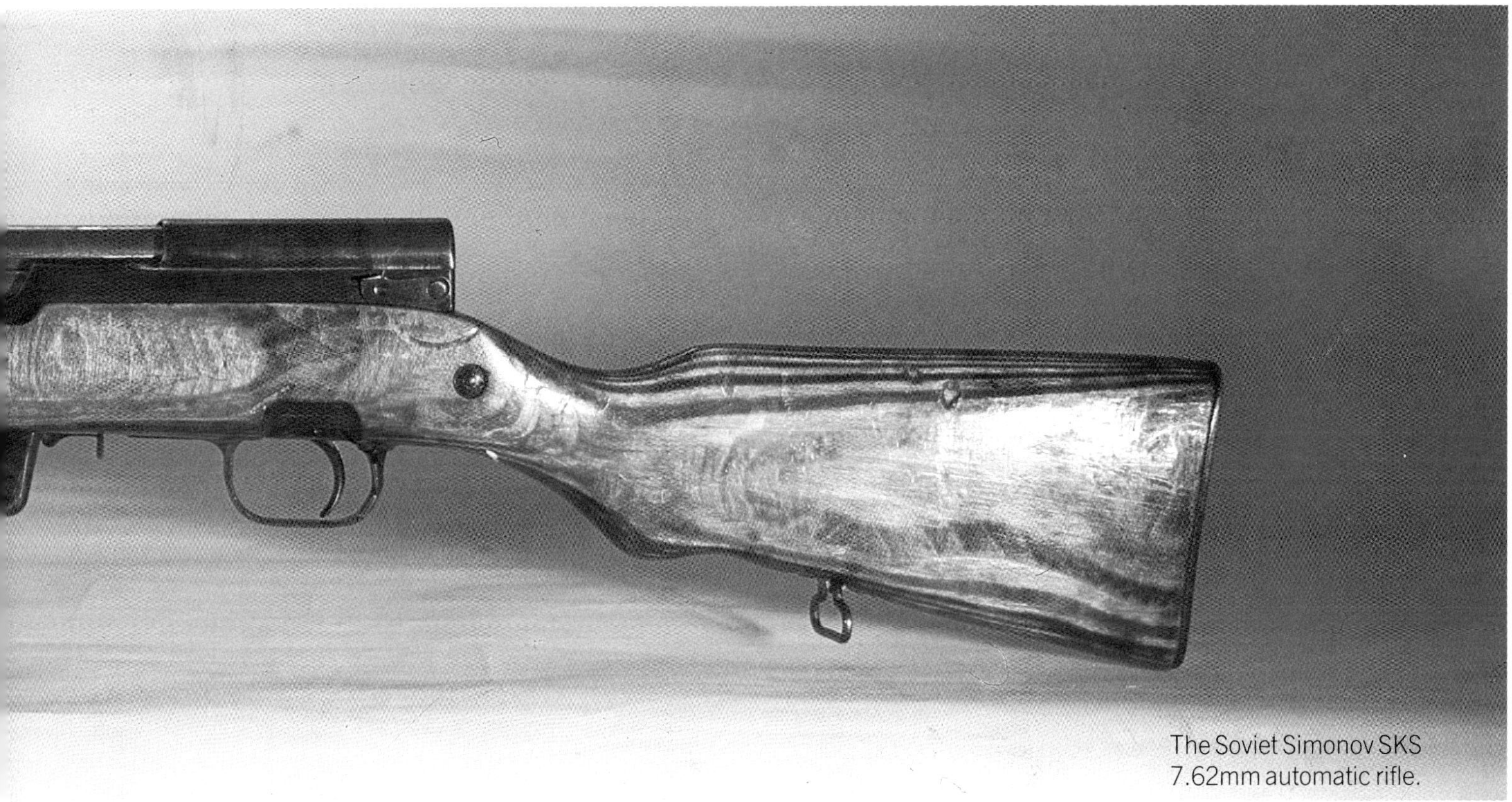

The Soviet Simonov SKS 7.62mm automatic rifle.

The SIG SG510 7.62mm automatic rifle with bipod mount.

The SIG SG542 7.62mm assault rifle stripped, showing the simplicity of the design.

be exported with the weapons by Norinco. In any case there are several Western world manufacturers making the Soviet cartridge by now, and there should be no difficulty in obtaining ammunition. Of the two the Simonov is probably the better hunting rifle, with a 20.5in barrel as against the 16.3in barrel of the Kalashnikov, which gives it the edge in accuracy and consistency.

Norinco also produce a line of .22 semi-automatic sporters; these are all the same basic weapon but with refinements – telescope sights of various grades, different barrel weights and lengths, adjustable stocks of varying qualities of finish, and even a take-down model in which the barrel can be quickly disconnected from the receiver for carriage or stowage. These appear to be well-finished and serviceable weapons which should be reasonably priced.

Other erstwhile military rifles available in semi-automatic sporting form include the SIG-AMT which is the civilianized version of the SIG 510, which in turn is derived from the Swiss Army's Sturmgewehr 57; the well-known Fabrique Nationale FN-FAL in FN-LAR form, the Colt AR-15, the Armalite AR-180, and sundry variations on the Garand design produced by Springfield Armory. Heckler & Koch market the HK91, which is the civil version of the West German Army's Gewehr 3 in 7.62mm NATO caliber and the HK93 which is the civil version of the 5.56mm HK33E. Heckler & Koch, though, take things one stage further by adapting the roller-locked breech mechanism of their military rifles to a sporting barrel and stock configuration to produce a handsome range of sporting weapons which benefit from the long and painstaking development and gradual improvement which has taken place in their military designs over the past twenty-odd years.

A new weapon with considerable civilian potential has recently been announced by the British Royal Small Arms Factory, the 'Enfield Cadet Rifle'. The British Army wanted a rifle for training cadets which would fire the new 5.56mm cartridge, but since it was to be solely for target shooting there was no need to burden the design with an automatic ability. So the Enfield XM-70 bull-pup has been modified by omitting the gas system, so converting it to a hand-operated magazine rifle; the firer simply grasps the cocking handle and moves it back and forward to eject and reload. The optical sight has been removed and a carrying handle with iron sight fitted, and such accessories as the grenade launcher and flash hider have been dispensed with. The result is a simple and effective rifle, firing the standard cartridge, and which cannot possibly be converted to automatic operation. Moreover there is a slip-in conversion kit which allows firing the .22 Long Rifle cartridge for indoor gallery practice. So far the Enfield factory has no plans for putting the Cadet Rifle on the civil market, but it cannot be denied that the package is a very attractive one.

In the purely civil field of rifles the bolt-action system has a considerable lead over the lever-action, the semi-automatic, the slide action, and the occasional single-shot action, and the Mauser bolt, with its forward locking lugs, is the unchallenged leader. The bolt action is pre-eminent for several reasons, some more apparent than real; it is familiar, safe, simple, and, in theory at least, the locking system which promises the best accuracy. This latter claim is based on the stiffness of a bolt-action receiver and on the tightness of a forward-lug system which does not permit any compression of the bolt (as does a rear-lug system) and thus minute setting-back of the cartridge case on firing. In practice it is doubtful if one shooter in ten thousand is good enough to gain advantage from this; military-caliber competition shooting in Britain was, for many years, shared between the Lee-Enfield

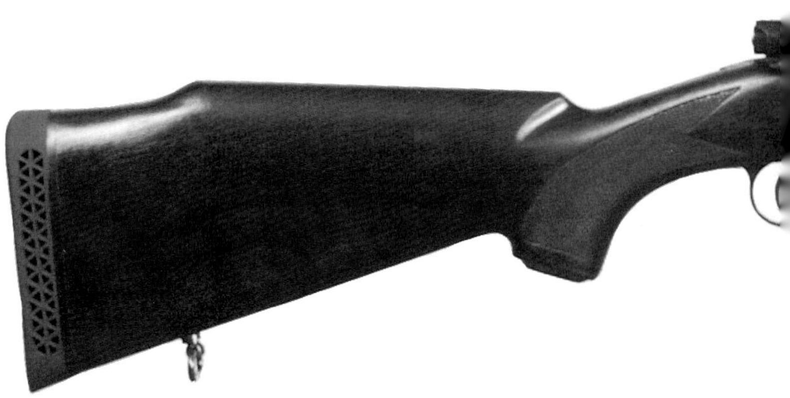

rifle, with its rear-locking bolt, and the Enfield Pattern 14 rifle with a forward-locking Mauser bolt, and honors were shared fairly equally between the two, results depending more upon the man behind the bolt than the bolt itself.

Slide action rifles can almost be counted on the fingers, and are rarely seen outside the USA. They have an advantage in handling; the firer's hands remain in the same position on the weapon, as he merely retracts the slide and pushes it forward, so that the second shot can be got off without having to reposition the hands and for rapidity of fire the slide action is probably the best of any manually-operated weapon. Since the internal mechanism is closely allied to that of a semi-auto there are those who complain that a slide gun will never compete in accuracy with a bolt action, but again this is a subjective judgement and depends upon the individual weapon and the firer. Certainly generations of shooters who cut their teeth on slide action .22s in years gone by found little to complain about.

Semi-automatic sporting rifles are usually either gas-operated or, in the smaller calibers, simple blowbacks. As noted above, the Heckler & Koch company have adapted their tried and proven roller-locked delayed blowback system to sporting rifles with success, but this is simply a matter of economics; it would be silly to ignore a system for which they were tooled up in order to develop something totally different for the sporting market. Auto-loaders have the advantage of being ready for the second shot almost instantly, without the firer having to shift his grip; they have the disadvantage, according to some critics, of being prone to jam at critical moments. On examination this latter fault can usually be laid at the door of hand-loaded ammunition; after all, military semi-automatics function reliably using similar mechanisms, so there is no inherent reason why a sporting semi-auto should go wrong provided it is supplied with factory ammunition, or at least ammunition up to factory standards.

Lever-action rifles are practically an American trademark, the few manufactured elsewhere in the world being mostly out-and-out copies of the familiar 'cowboy carbine' Winchester design. In general, lever-action rifles are popular for short-range hunting in close country where game and shooter usually come face-to-face at ranges less than 100 yards. For this reason the lever-action is usually a light weapon with sufficient accuracy for the task in hand but with little in reserve for long range shooting. Critics frequently condemn the lever-action for being less accurate than a bolt action, overlooking their fundamental purpose. The light barrel and receiver of a lever action rifle does dispose it to flexing and hence to rather less accuracy than a stiffer and heavier action, but, again, the lever-action would not sell if it was an inaccurate system, and in general the lever action rifle is well capable of the degree of accuracy which its employment demands.

A criticism levelled at lever-actions in the past has been that they generally fired low-power cartridges such as the .30-30, .32-20 and various pistol cartridges. This is no longer a valid argument, since modern lever-actions can be chambered for such potent loadings as the .250 Savage, .243 Winchester, .308 Winchester and .375 Winchester.

Whatever may be currently fashionable, or easy, or convenient, you can be sure that somewhere there are people who prefer to do things differently. Most photographers go out with a 35mm camera and shoot a 36-exposure roll, trusting that one shot will be the one they want. At the other end of the scale is the man who packs an 8×10 plate camera into the mountains, spends all day seeking the best vantage point, and comes back with just one picture, carefully composed and exposed. In similar vein there are still shooters who prefer to go out into the woods with a single-shot rifle and a handful of cartridges, intent on making every shot count, and there are still manufacturers willing to supply the weapon they desire. In center-fire calibers the most usual type of single-shot is the falling-block breech pattern in which an under-lever or side-lever slides the breech-block downwards like that of an artillery piece. This design gives a short and extremely rigid receiver and produces rifles capable of the utmost accuracy which can handle the most powerful cartridges. The alternative is the break-open system, like that more commonly used in shotguns, which is generally to be found with the less powerful cartridges. In .22 rimfire caliber the bolt action is practically the only system used, since it is simple to manufacture and amply strong for the cartridge.

For those who desire to be different and perhaps to savor the aristocratic style, the double rifle can still be found. The classic double, favored by big game hunters, is the side-by-side which is barely distinguishable from a shotgun at first glance, but there is also the over-and-under configuration for those who prefer it. Either way the weapon is going to be a hand-made – or at least hand-finished – beauty with a price-tag to match; $4000 is the opening bid in this game and the top limit is whatever the customer cares to specify in the way of fit and finish. The double rifle has a feeling all of its own, and may take some getting used to, but the accuracy is rarely in question. One point frequently overlooked by those new to these guns is that they must be carefully built to match the

The traditional lines of the Whitworth Sporting Rifle, available in 7mm Remington Magnum, .300 Winchester Magnum, .375 Holland and Holland and .458 Winchester, all very powerful loadings.

ballistics of the chosen cartridge, particularly in the side-by-side version; there is one set of sights for two barrels, and thus the barrels must converge slightly to place both bullets on the point of aim at a specified distance. This distance depends upon what the rifle is to be used for and what the cartridge is going to be; and if it has been 'regulated' for, say, a 180 grain bullet, then don't expect the same result if you decide to change to a 200-grain bullet. To put it bluntly, double rifles are for people who know what they are doing.

Competition rifles, like competition pistols, come in all shapes and sizes to suit the various types of contest and the varying skills of the shooters. From the club member at the lower end of the scale to the Olympic contestant at the top, there are rifles to suit at commensurate prices. In this field the bolt action reigns supreme and the price asked reflects the degree of precision in assembly, the quality of finish and the accessories, notably the sights.

The .22 Long Rifle cartridge is the international standard for 'small-bore' match shooting; for center-fire matches there is a wider choice of cartridge, though military calibers are often national standards and the .308 Winchester (7.62mm NATO) is another universal round. Generally speaking a match rifle will have a heavier barrel than a sporting rifle, will have the bore more carefully rifled and honed to a perfect finish, will be fitted with infinitely adjustable aperture sights (both front and rear) and will have a stock which is heavier and offers a better grip than a sporting stock. Many target stocks are adjustable, either by having the butt-plate and cheek-piece mounted on arms which can be slid back and forth or by having them removeable and offering spacers to fill them out to the desired position. One or two highly-specialised rifles are offering 'electronic' (more properly 'electrical') operation of the firing mechanism; the firing pin is part of a solenoid controlled by a micro-switch under the trigger, so that the slightest touch is sufficient to propel the firing pin forward and a simple circuit-breaking switch offers positive safety. The object behind this is to reduce the 'lock time' to the absolute minimum. Lock time is the time which elapses between the trigger releasing the firing impulse and the bullet leaving the bore, and the shorter it is, the less danger there is of involuntary movement by the firer during the passage of the bullet up the bore. The average bolt action rifle has a fairly long lock time; the shortest mechanically-actuated system is the Martini falling-block, which has a short firing pin inside the block and thus has a very short pin travel, and Martini blocks were popular for many years, and still are in Europe, for .22 match rifles.

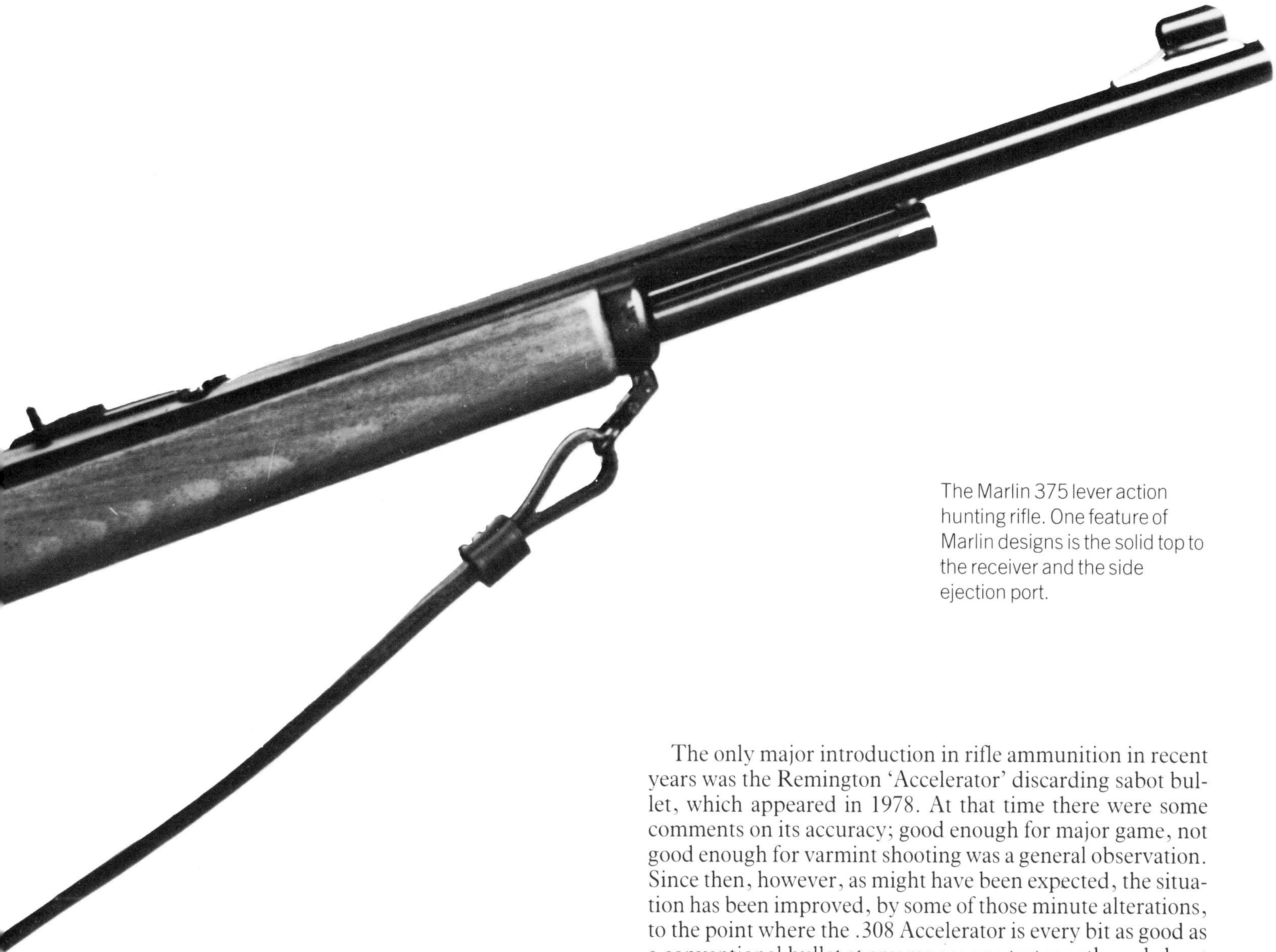

The Marlin 375 lever action hunting rifle. One feature of Marlin designs is the solid top to the receiver and the side ejection port.

Rifle ammunition makes haste slowly; manufacturers put a cartridge on the market only after considerable testing, and thus the final product generally behaves well and requires little adjustment thereafter. Then somebody thinks about changing the bullet, for more velocity or better kills or whatever, and, after more testing, different bullets gradually appear. But to expect any sensational changes in rifle ammunition is to court disappointment. Indeed, most of the changes are virtually invisible, being such matters as improvements on the design and manufacture of bullets so as to improve their expansion or minute changes in contour to improve flight characteristics.

The only major introduction in rifle ammunition in recent years was the Remington 'Accelerator' discarding sabot bullet, which appeared in 1978. At that time there were some comments on its accuracy; good enough for major game, not good enough for varmint shooting was a general observation. Since then, however, as might have been expected, the situation has been improved, by some of those minute alterations, to the point where the .308 Accelerator is every bit as good as a conventional bullet at any range; one tester enthused about ⅝ inch groups, though he omitted to say at what range he made them.

At the time of the introduction of the 'Accelerator' rounds I observed that they ought to have a useful military application in the armor-piercing field; at that time we were all unaware that, in fact, this was exactly what was going on behind the scenes, and it was not until late 1983 that the US Army revealed its 'SLAP' cartridges. SLAP means 'Saboted Light Armor Penetrator' and in 7.62mm caliber is virtually the .308 Accelerator with an armor-piercing sub-projectile instead of one resembling a conventional jacketed bullet. A particular advantage of the discarding sabot technique is that the sub-projectile can be designed to have specific characteristics, in this case with the long spire-pointed nose designed primarily for piercing. Both spin-stabilized and fin-stabilized sub-projectiles have been developed, the latter being long dart-like penetrators, though fin-stabilization appears to be better suited to .50 caliber cartridges for use in machine guns.

SPECIFICATIONS OF REPRESENTATIVE MODERN RIFLES

NAME	CALIBER	OPERATION	LENGTH	WEIGHT EMPTY	BARREL LENGTH	MAGAZINE CAPACITY	REMARKS
ANSCHUTZ: J. G. Anschutz GmbH, D-7900 Ulm/Donau, West Germany							
Model 1411	.22LR	Bolt	46.0in	11.0lbs	27.5in	S/S	Match rifle
Model 64MS	.22LR	Bolt	39.5in	8.0lbs	21.8in	S/S	Match rifle
ARMSPORT: Armsport Inc, Miami FL 33142							
Emperor 4000	Various	Double	Dimensions to customer's specification. Side-by-side or over/under				
BERETTA: Pietro Beretta Spa, Gardone Val Trompia, Brescia, Italy							
BM59	7.62mm	Gas	43.1in	10.1lbs	19.3in	20	Garand action
AR70	5.56mm	Gas	37.6in	8.4lbs	17.7in	30	Assault rifle
SC70	5.56mm	Gas	37.8in	8.5lbs	17.7in	30	Folding butt
SC70 Short	5.56mm	Gas	32.3in	8.21lbs	12.6in	30	Carbine model
BINGHAM: Bingham Ltd., Norcress GA 30093							
AK-22	.22LR	Blowback	35.5in	6.1lbs	17.5in	15, 28	Replica of AK47

NAME	CALIBER	OPERATION	LENGTH	WEIGHT EMPTY	BARREL LENGTH	MAGAZINE CAPACITY	REMARKS
BROWNING: Browning Arms, Morgan, UT 84050							
Model '78	Various	Block	42.0in	8.5lbs	26.0in	S/S	No iron sights
Model 1895	.30-06	Lever	42.0in	8.0lbs	24.0in		
B-78	.45-70	Block	40.7in	8.7lbs	24.0in	S/S	
BAR	.308	Gas	43.0in	7.4lbs	22.0in	4	
BAR22	.22LR	Blowback	38.2in	6.2lbs	20.2in	15	Tubular mag.
BPR22	.22LR	Slide	38.2in	6.2lbs	20.2in	11 or 15	Tubular mag.
BL-22	.22RF	Lever	36.7in	5.0lbs	20.0in	15-22	Tubular mag
B-92	.44RemMag	Lever	37.5in	5.5lbs	20.0in	11	Tubular mag.
BLR	.308	Lever	39.8in	6.9lbs	20.0in	4	Also .243, .358
BBR	.300WMag	Bolt	44.5in	8.0lbs	24.0in	3	Other calibers

The Beretta SC70 Short assault rifle.

NAME	CALIBER	OPERATION	LENGTH	WEIGHT EMPTY	BARREL LENGTH	MAGAZINE CAPACITY	REMARKS
B.S.A.: Birmingham Small Arms Co., Birmingham, England							
CF-2	Various	Bolt	45.0in	7.8lbs	24.0in	3	
CFT	.308	Bolt	47.0in	10.8lbs	26.5in	S/S	
Martini ISU	.22LR	Block	44.0in	10.8lbs	28.0in	S/S	Martini action
CETME: Empresa Nacional 'Santa Barbara' de Industrias Militares SA, Madrid 10, Spain							
Model C	7.62mm	Delay B/B	40.0in	9.3lbs	17.7in	20	Military
Model L	5.56mm	Delay B/B	36.4in	7.9lbs	15.7in	30	Military
CZ: Ceskoslovenska Zbrojovka, Uhersky Brod, Czechoslovakia							
vz/58	7.62mm	Gas	33.2in	6.9lbs	15.8in	30	Czech military
ENFIELD: Royal Small Arms Factory, Enfield Lock, Middlesex, England							
XM-70	5.56mm	Gas	30.3in	8.6lbs	20.4in	30	Military bullpup
L39A1	7.62mm	Bolt	46.5in	9.71lbs	27.6in	10	Military target
L42A1	7.62mm	Bolt	46.5in	9.8lbs	27.5in	10	Sniper

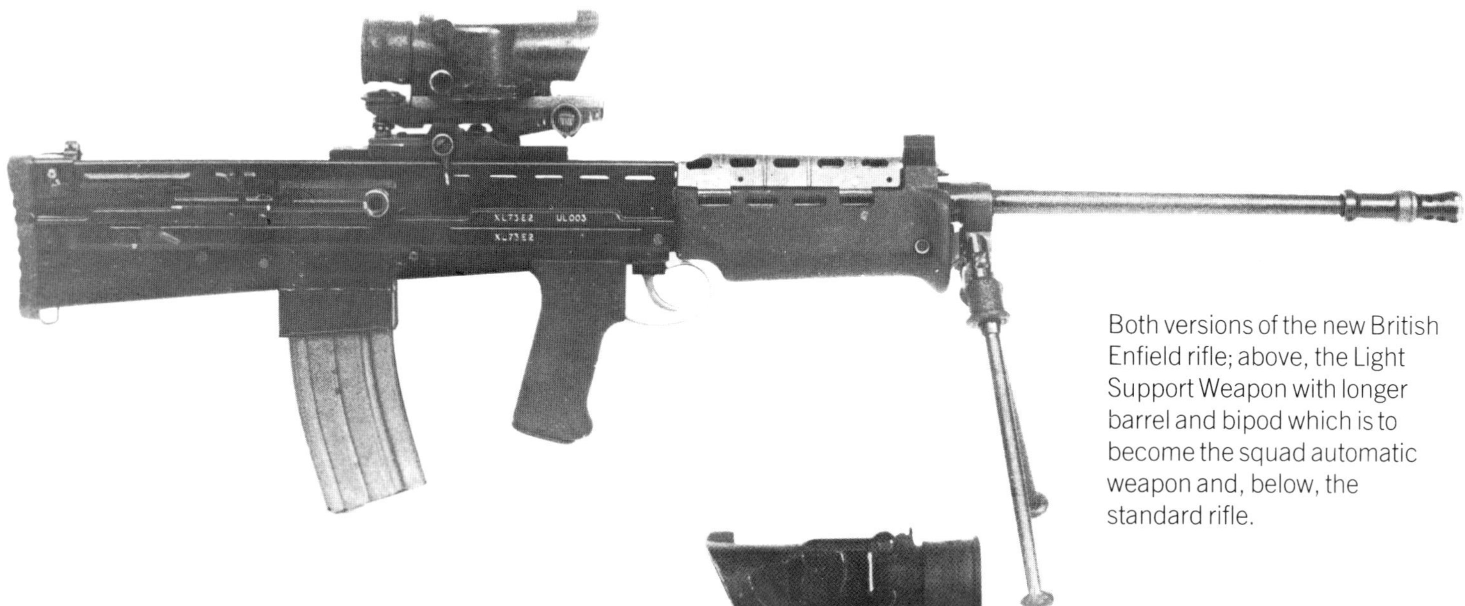

Both versions of the new British Enfield rifle; above, the Light Support Weapon with longer barrel and bipod which is to become the squad automatic weapon and, below, the standard rifle.

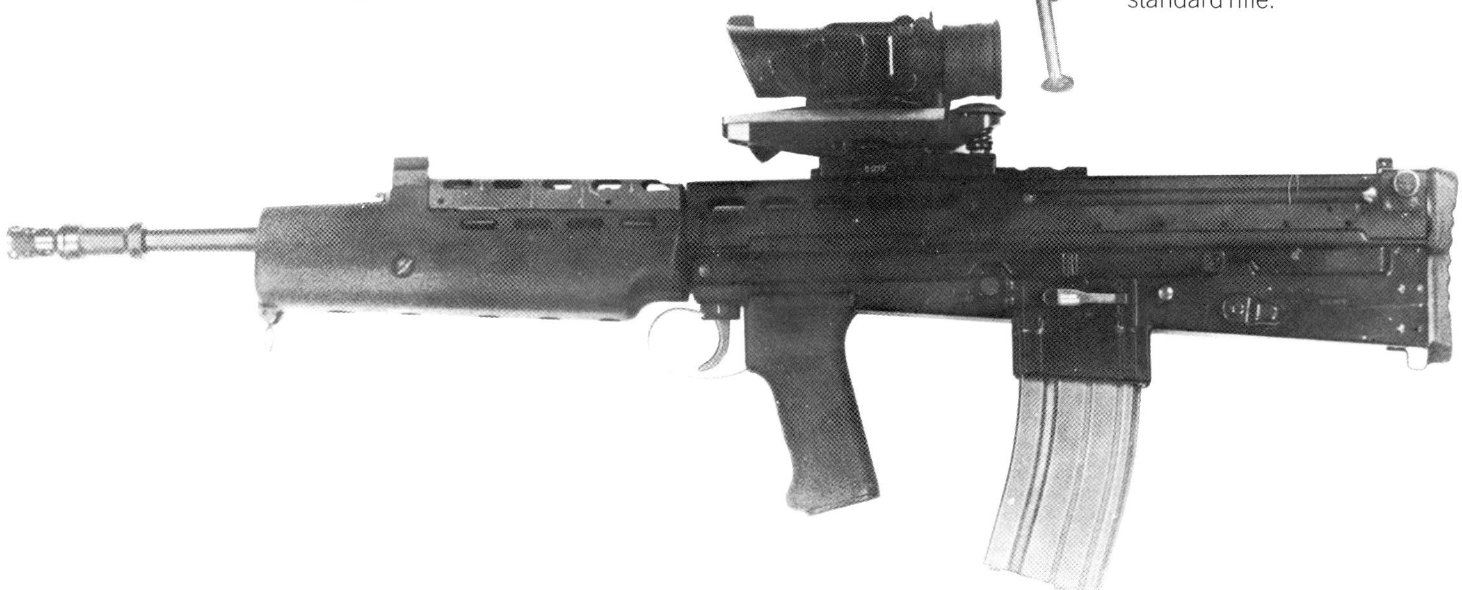

NAME	CALIBER	OPERATION	LENGTH	WEIGHT EMPTY	BARREL LENGTH	MAGAZINE CAPACITY	REMARKS
ENFIELD Contd.							
Envoy	7.62mm	Bolt	46.7in	10.5lbs	27.5in	10	Target
Enforcer	7.62mm	Bolt	47.5in	10.5lbs	27.5in	10	Police sniper
ERMA: Erma-Werke GmbH, Postbox 1269, D-8060 Munich-Dachau, West Germany							
ESG-22	.22LR	Blowback	35.5in	6.0lbs	18.0in	15	Also .22WMRF with 12-shot mag.
EG-73	.22WMRF	Lever	37.4in	6.0lbs	19.3in	12	Tubular mag.
FABRIQUE NATIONAL: Fabrique National Herstal SA, B-4400 Herstal, Belium							
FAL	7.62mm	Gas	41.5in	9.5lbs	21.0in	20	Several variants
FNC	5.56mm	Gas	38.9in	8.5lbs	17.7in	30	Folding butt
LAR	7.62mm	Gas	44.5in	9.5lbs	21.0in	20	Sporter FAL
Model 30-11	7.62mm	Bolt	44.0in	11.1lbs	19.8in	5	Sniper
FEINWERKBAU: Feinwerkbau Westinger & Altenburger KG, D-7238 Oberndorf-a-Neckar, West Germany							
Model 2000	.22LR	Bolt	43.8in	9.7lbs	26.3in	S/S	Match rifle
FERLACH: Franz Sodia Jagdgewehrfabrik, A-9170, Ferlach, Austria							
Double Rifle	Any	Double	All dimensions to suit customer. Side-by-side				
GALIL: Israeli Military Industries, Ramat Ha Sharon, Israel							
Galil AR	5.56mm	Gas	38.5in	9.3lbs	18.1in	35	Assault rifle
Galil SAR	5.56mm	Gas	33.5in	8.0lbs	13.1in	35	Carbine model
Sniper	7.62mm	Gas	43.9in	14.1lbs	20.0in	20	
HARRINGTON & RICHARDSON: Harrington & Richardson, Gardner, MA 01440							
Model 155	.45-70	Break	39.0in	7.3lbs	24.0in	S/S	
Model 5200	.22LR	Bolt	46.0in	11.0lbs	28.0in	S/S	Target rifle
Model 700	.22LR	Blowback	43.2in	6.5lbs	22.0in	5	
Model 750	.22RF	Bolt	39.0in	5.0lbs	22.0in	S/S	
Model 865	.22RF	Bolt	39.0in	5.0lbs	22.0in	5	
HECKLER & KOCH: Heckler & Koch GmbH, D-7238 Oberndorf-a-Neckar, West Germany							
Gewehr 3A3	7.62mm	Delay B/B	40.0in	9.4lbs	17.8in	20	Military. Several variant models
HK33E	5.56mm	Delay B/B	37.0in	8.1lbs	15.4in	25 version	Also carbine
HK91	.308	Delay B/B	40.3in	9.5lbs	19.0in	5, 20	Sporter G3
HK93	.223	Delay B/B	35.5in	7.8lbs	16.1in	5, 20	Sporter HK33E

NAME	CALIBER	OPERATION	LENGTH	WEIGHT EMPTY	BARREL LENGTH	MAGAZINE CAPACITY	REMARKS
HECKLER & KOCH Contd.							
G-11	4.7mm	Gas	29.5in	7.9lbs	21.3in	50	Caseless ammo.
G-41	5.56mm	Delay B/B	39.3in	8.6kg	18.9in	30	3-round burst
PSG-1	7.62mm	Delay B/B	47.6in	17.8lbs	25.6in	20	Sniper
Model 940	.30-06	Delay B/B	47.2in	8.6lbs	21.6in	3 or 10	
Model 300	.22LR	Blowback	39.5in	5.7lbs	19.7in	5	
Model 770	.308	Delay B/B	42.8in	7.5lbs	19.6in	3	Sporter

The Heckler & Koch Model 300 sporting rifle.

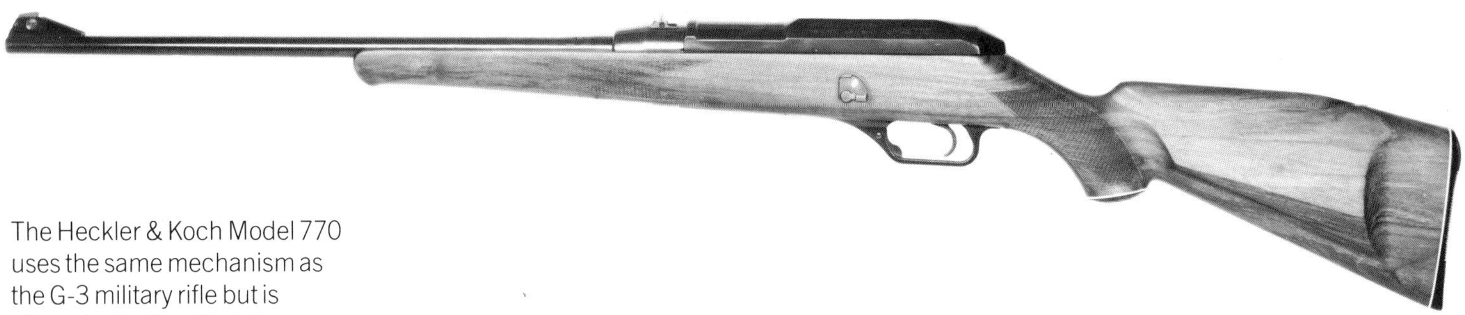

The Heckler & Koch Model 770 uses the same mechanism as the G-3 military rifle but is otherwise built entirely for sporting use.

The Heckler & Koch HK33E 5.56mm assault rifle, fitted with a Litton night sight.

NAME	CALIBER	OPERATION	LENGTH	WEIGHT EMPTY	BARREL LENGTH	MAGAZINE CAPACITY	REMARKS
HEYM: F.W. Heym, D-8732, Munnerstadt, West Germany							
Model 88B	Various	Double	42.0in	7.5lbs	25.0in	–	Side-by-side
Model 55B	Various	Double	42.0in	8.0lbs	25.0in	–	Over/under
Model SR20	Various	Bolt		7.5lbs	24.0in	S/S	
Model HR30	Various	Block		7.0lbs	24.0in	S/S	
IVER JOHNSON: Iver Johnson Arms Inc., Middlesex, NJ 08846							
Plainfield Carbine	.30 Carb	Gas	35.5in	6.5lbs	18.0in	15	Replica M1 Carbine
JAPANESE SERVICE: Howa Machinery Ltd, Shinkawacho, Nagoya, Japan							
Type 64	7.62mm	Gas	39.0in	9.7lbs	17.7in	20	Military
KIMBER: Kimber of Oregon Inc., Clackamas, OR 97015							
Model 82	.22RF	Bolt	41.0in	6.3lbs	24.0in	5	
Model 84	.223Rem	Bolt	40.5in	6.5lbs	22.5in	5	
KLEINGUNTHER: Kleingunther's, Seguin, TX 78155							
K-15	Various	Bolt	43.5in	7.8lbs	24.0in	3	
KRICO: Kriegeskorte & Co GmbH, D-7000 Stuttgart 61, West Germany							
Model 302	.22LR	Bolt	42.5in	6.5lbs	24.0in	5 or 10	
Model 400	.22Hor	Bolt		6.6lbs	24.0in	5	
Model 640S	.223	Bolt		7.5lbs	20.0in	5	Other calibers
Model 650S	.223	Bolt	46.0in	10.6lbs	26.0in	5	Target rifle
Model 720LR	7mm	Bolt	41.2in	6.9lbs	21.0in	3	Sporting

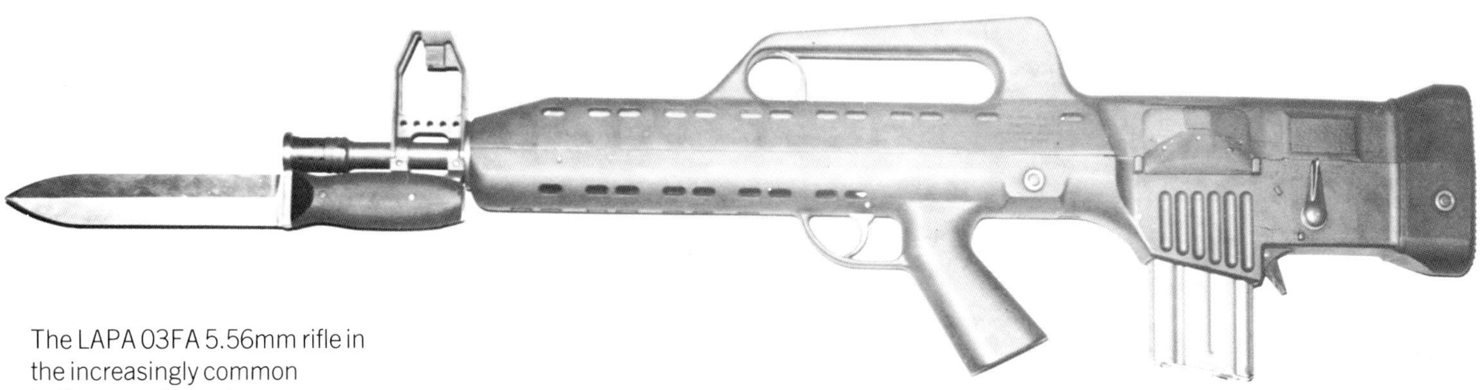

The LAPA 03FA 5.56mm rifle in the increasingly common bullpup configuration.

NAME	CALIBER	OPERATION	LENGTH	WEIGHT EMPTY	BARREL LENGTH	MAGAZINE CAPACITY	REMARKS
LAPA: Laboratorio de Pesquisas de Armamento Automatico Lda., 20510 Rio de Janiero, Brazil							
Model 03FA	5.56mm	Gas	29.0in	7.7lbs	19.25in	20, 30 or 40	Bullpup; under development

NAME	CALIBER	OPERATION	LENGTH	WEIGHT EMPTY	BARREL LENGTH	MAGAZINE CAPACITY	REMARKS
MARLIN: Marlin Firearms Co., New Haven, CT 06473							
Model 15Y	.22RF	Bolt	33.0in	4.7lbs	16.25in	S/S	
Model 39A	.22RF	Lever	40.0in	6.5lbs	24.0in	19-26	Tubular mag.
Model 336C	.30-30	Lever	38.5in	7.0lbs	20.0in	6	Tubular mag.
Model 444S	.444	Lever	40.5in	7.5lbs	22.0in	4	Tubular mag.
Model 780	.22RF	Bolt	41.0in	5.5lbs	22.0in	7	
Model 990	.22RF	Blowback	40.7in	5.5lbs	22.0in	18	Tubular mag.
Model 1894	.44Mag	Lever	37.5in	6.0lbs	20.0in	10	Tubular mag.
Glenfield 60	.22RF	Blowback	41.0in	5.5lbs	22.0in	18	Tubular mag.
Glenfield 75C		Blowback	36.7in	5.0lbs	18.0in	14	Tubular mag.
MAS: Manufacture Nationale d'Armes de St Etienne, St Etienne, France							
MAS 49	7.5mm	Gas	43.5in	10.0lbs	22.75in	10	Military
FA-MAS	5.56mm	Delayed Blowback	29.8in	8.2lbs	19.2in	25	Military
MAUSER: Mauser-Werke Oberndorf GmbH, D-7238 Oberndorf-a-Neckar, West Germany							
SP-66	7.62mm	Gas		7.9lbs	26.8in	3	Sniper
Model 66	.308	Bolt	41.0in	7.5lbs	24.0in	5	Other calibers
Model 77	.308	Bolt	44.0in	7.5lbs	24.0in	3	Other calibers

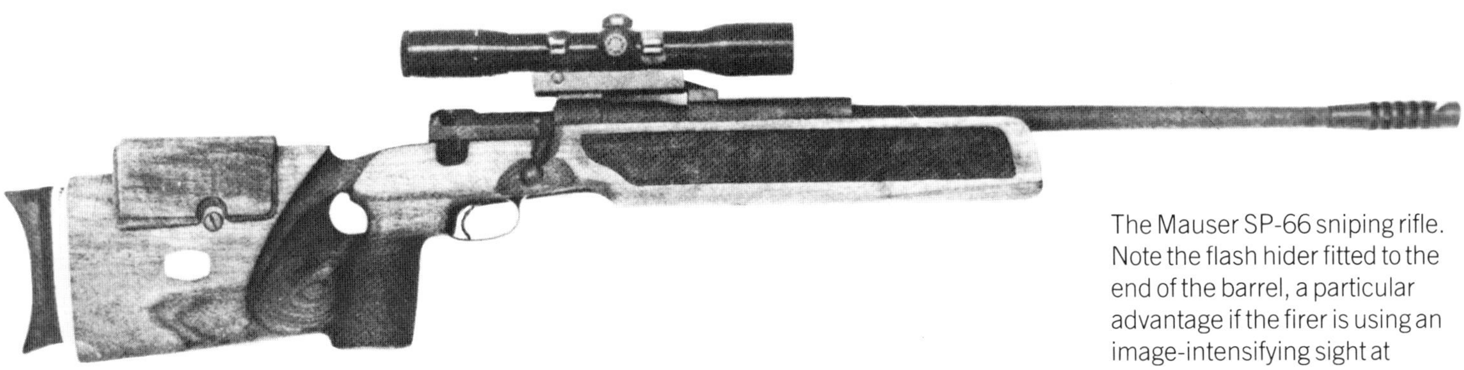

The Mauser SP-66 sniping rifle. Note the flash hider fitted to the end of the barrel, a particular advantage if the firer is using an image-intensifying sight at night.

NAME	CALIBER	OPERATION	LENGTH	WEIGHT EMPTY	BARREL LENGTH	MAGAZINE CAPACITY	REMARKS
MOSSBERG: O.F. Mossberg, North Haven, CT 06473							
Model 144	.22LR	Bolt	43.0in	8.0lbs	27.0in	7	Target rifle
Model 341	.22RF	Bolt	43.5in	5.5lbs	24.0in	7	
Model 353	.22LR	Blowback	38.0in	5.0lbs	18.0in	7	
Model 380	.22LR	Blowback	40.0in	5.5lbs	20.0in	15	Tubular mag.

NAME	CALIBER	OPERATION	LENGTH	WEIGHT EMPTY	BARREL LENGTH	MAGAZINE CAPACITY	REMARKS
NORINCO: China North Industries Corp., Beijing, China							
Type 56	7.62mm	Gas	40.2in	8.5lbs	20.5in	5 or 10	Simonov copy
Type 56S	7.62mm	Gas	35.3in	9.5lbs	16.3in	30	Kalashnikov copy
Type 68	7.62mm	Gas	40.5in	7.7lbs	20.5in	15	Military
EM751	.22LR	Bolt	51.2in	14.3lbs	29.5in	S/S	Match rifle
EM752	.22LR	Bolt	44.6in	10.6lbs	27.2in	S/S	Match rifle
EM763	.22LR	Bolt	38.8in	11.0lbs	21.6in	S/S	Target rifle
JW-9	.22LR	Blowback	41.9in	6.8lbs	22.8in	5	Sporter
JW-12	22LR	Blowback	40.3in	6.17lbs	21.5in	15	Take-down
OVM: Olympio Viera de Mello, Rio de Janiero, Brazil							
Assault carbine	5.56mm	Gas	36.6in		18.3in	20 or 30	Development
PARKER-HALE: Parker-Hale Ltd., Golden Hillock Road, Birmingham, England							
Model 82	7.62mm	Bolt	45.7in	10.6lbs	26.0in	5	Military sniper
Model 83	7.62mm	Bolt	46.7in	11.0lbs	26.0in	S/S	Cadet rifle L18A1
Super 1200	Various	Bolt	45.0in	7.3lbs	24.0in	3	
REMINGTON: Remington Arms Co., Bridgeport CT 06602							
Model 4	.308	Gas	42.0in	7.5lbs	22.0in	4	Also other calibers
Model Six	.308	Slide	42.0in	7.5lbs	22.0in	4	And other calibers
Model 40XB	Various	Bolt	47.0in	9.3lbs	27.3in	S/S	Match rifle
Model 40XC	7.62mm	Bolt	42.5in	10.0lbs	23.3in	5	National Match
Model 40XR	.22LR	Bolt	43.0in	10.0lbs	24.0in	S/S	Match rifle
Model 66MB	.22LR	Blowback	38.5in	4.0lbs	19.7in	14	Tubular mag., nylon stock
Model 541S	.22RF	Bolt	42.7in	5.5lbs	24.0in	5	
Model 552A	.22RF	Blowback	40.0in	5.7lbs	21.0in	15-20	Tubular mag.
Model 572 Fieldmaster	.22RF	Slide	42.0in	5.5lbs	21.0in	14-20	Tubular mag.
Model 700	.22-250	Bolt	43.5in	7.0lbs	22.0in	3	Also 24in bbl
Model 788	.223	Bolt	41.7in	7.5lbs	18.5in	3	Also 24in bbl
ROSSI: Amadeo Rossi SA, Postbox 28, São Leopoldo, Brazil							
Model 62SA	.22RF	Slide	39.3in	5.8lbs	23.0in	14-20	Tubular mag.

NAME	CALIBER	OPERATION	LENGTH	WEIGHT EMPTY	BARREL LENGTH	MAGAZINE CAPACITY	REMARKS
RUGER: Sturm Ruger & Co., Southport, CT 06490							
Model 1	Various	Block	42.0in	8.2lbs	26.0in	S/S	Also 22, 24in bbls
Model 3 Carbine	.375Win	Block	38.5in	6.0lbs	22.0in	S/S	Other calibers

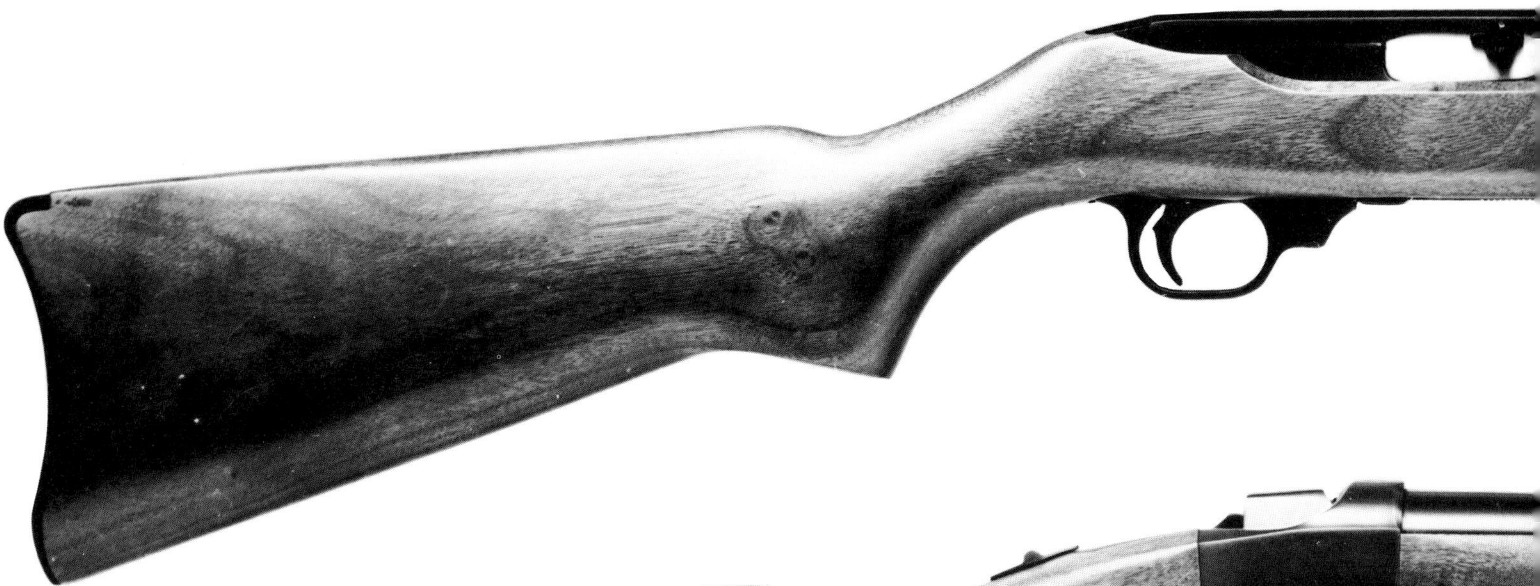

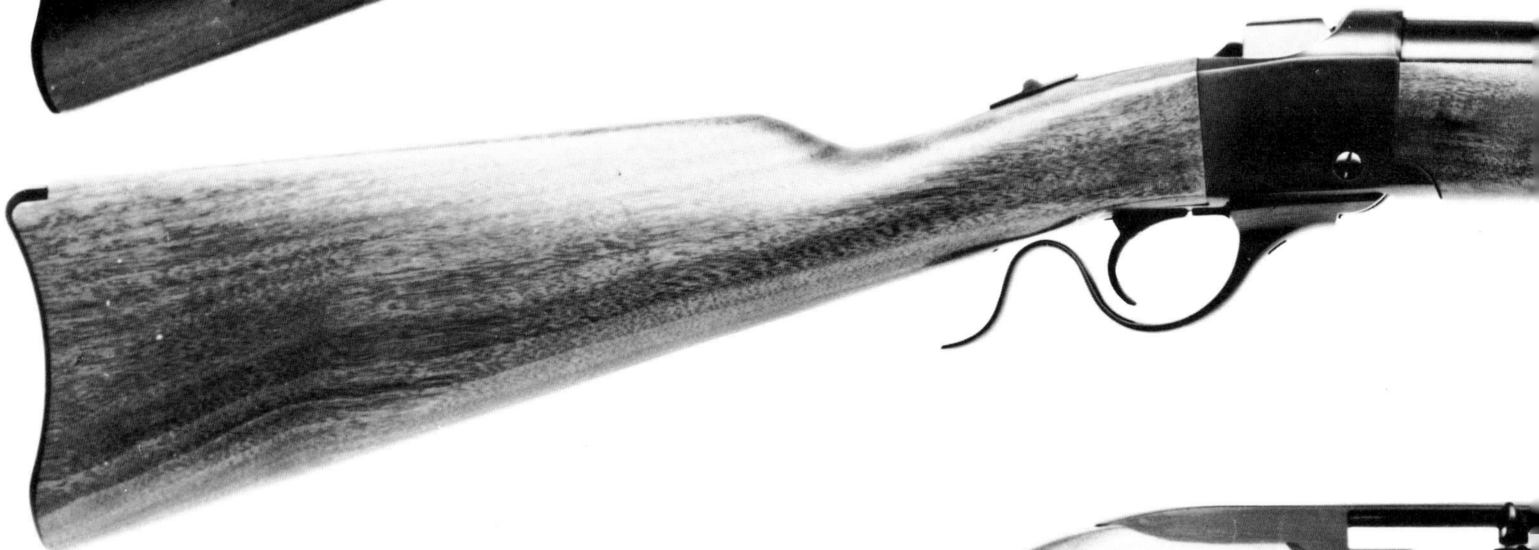

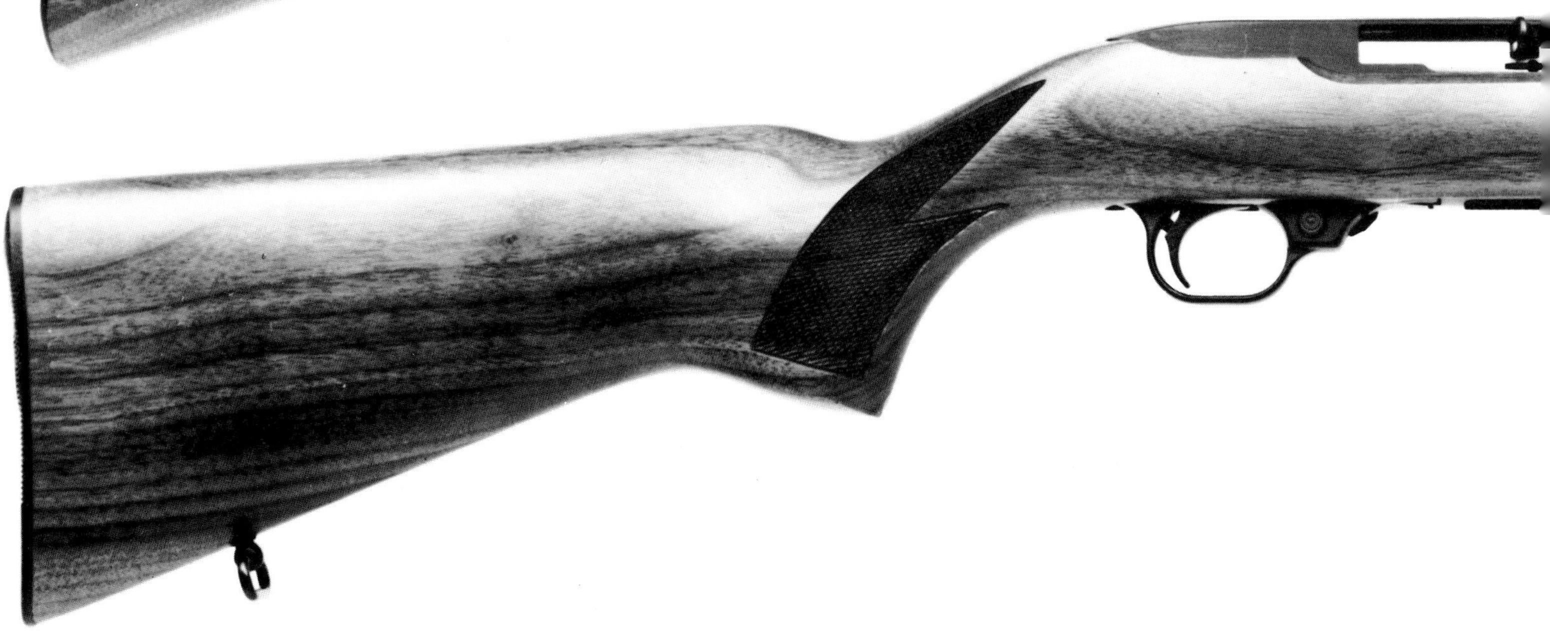

NAME	CALIBER	OPERATION	LENGTH	WEIGHT EMPTY	BARREL LENGTH	MAGAZINE CAPACITY	REMARKS
RUGER Contd.							
Model 10/22	.22LR	Blowback	36.7in	5.8lbs	18.5in	10	Rotary mag.
Model 44	.44Mag	Gas	36.3in	5.8lbs	18.5in	4	Carbine

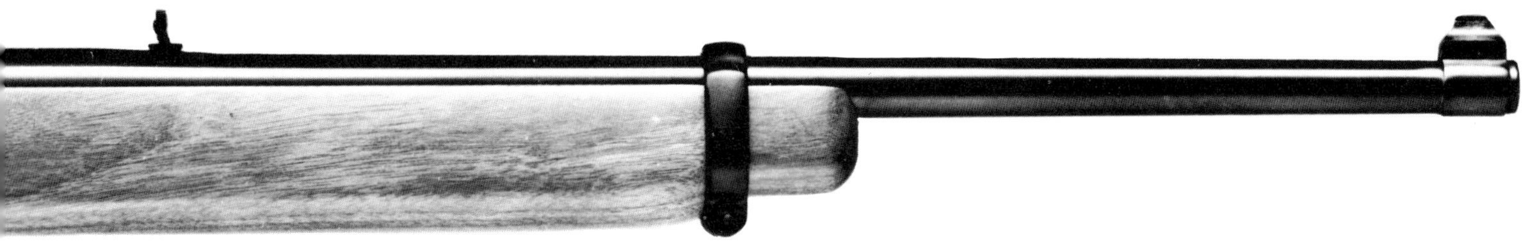

Above: The powerful Ruger Model 44 Carbine.

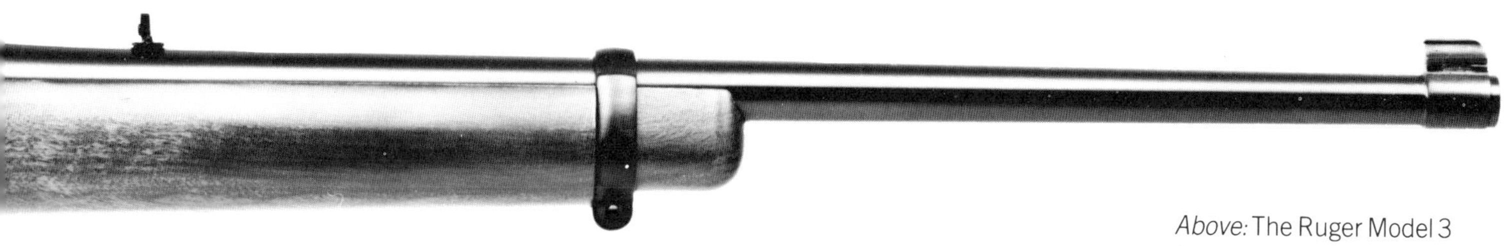

Above: The Ruger Model 3 Single-Shot Carbine, available in various calibers including .223, .44 Magnum and .45-70.

Above: The Ruger 10/22 Deluxe Model Sporter Rifle.

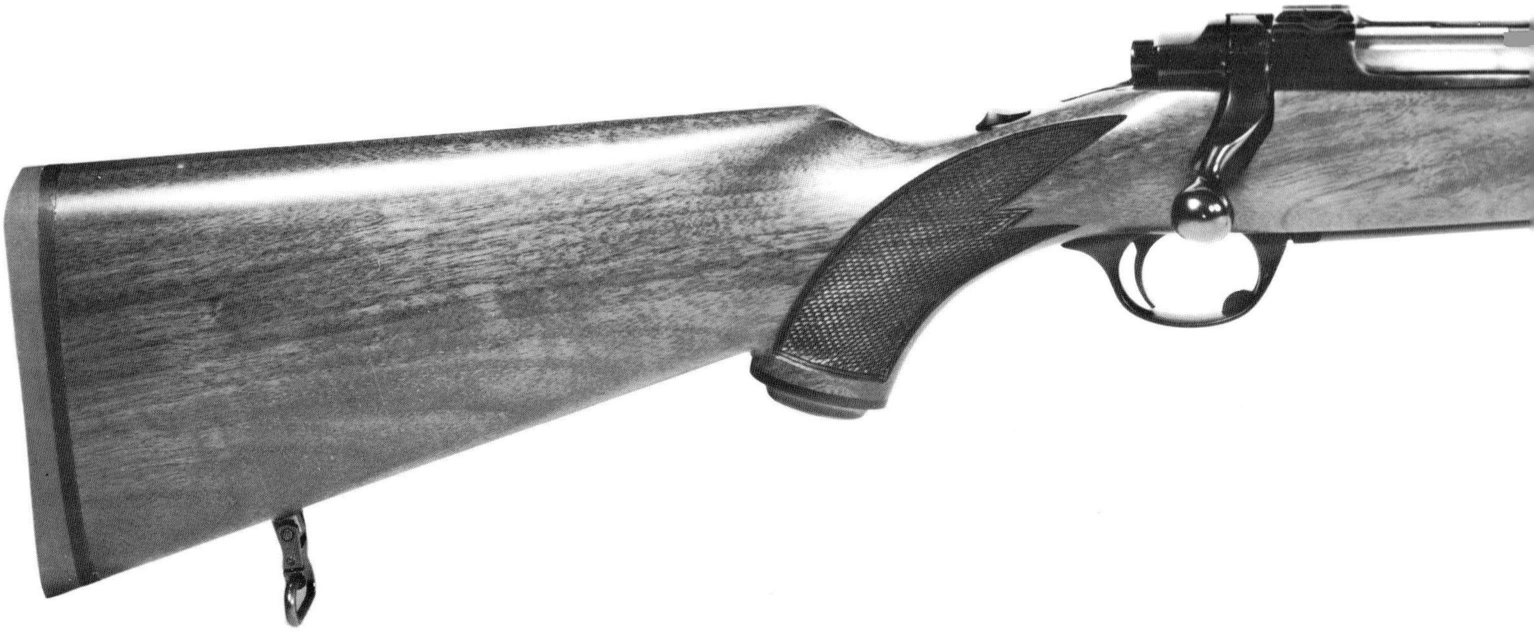

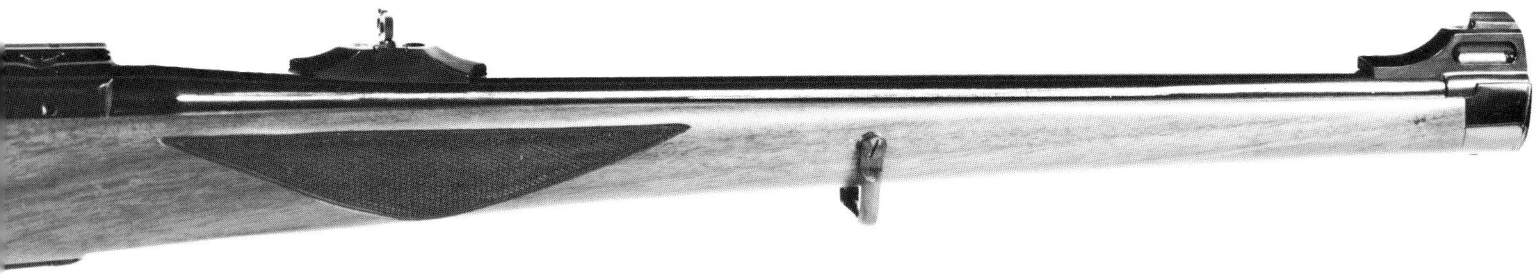

Above: The Ruger Model 77 with Mannlicher stock.

Above: The lighter 'Tropical' version of the Ruger Model 77.

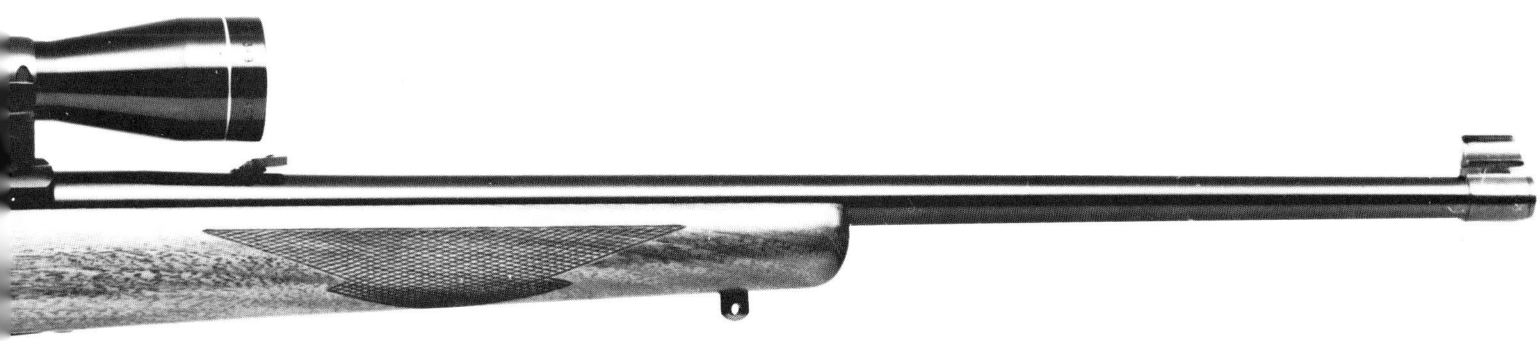

Above: The Ruger Model 77/22 incorporates integral scope mounts.

NAME	CALIBER	OPERATION	LENGTH	WEIGHT EMPTY	BARREL LENGTH	MAGAZINE CAPACITY	REMARKS
RUGER Contd.							
Model 77	.308	Bolt	42.0in	7.1lbs	18.5in		Also .243, .250/3000
Model 77 Ultra-light	.308	Bolt	40.0in	6.2lbs	20.0in	5	Also .257, .270, .30-06 etc.

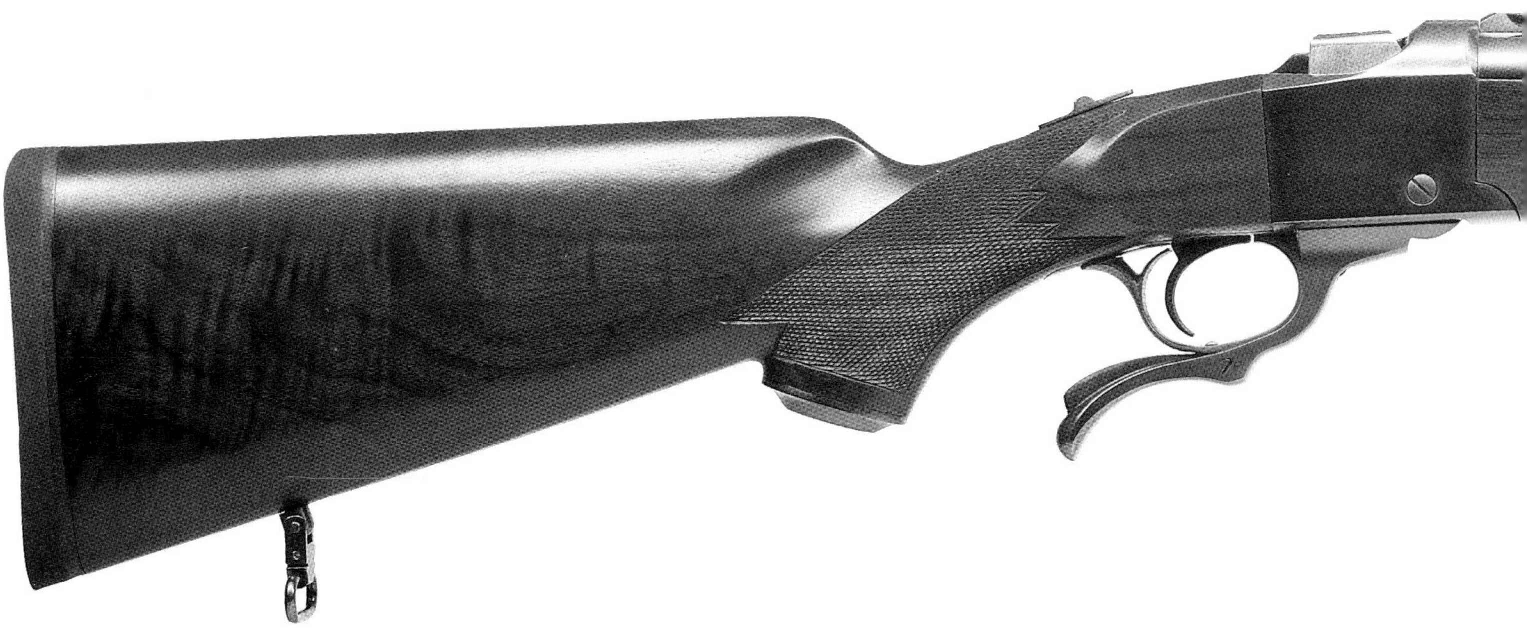

NAME	CALIBER	OPERATION	LENGTH	WEIGHT EMPTY	BARREL LENGTH	MAGAZINE CAPACITY	REMARKS
RUGER Contd.							
Model 77/22	.22LR	Bolt	39.25in	5.8lbs	20.0in	10	
Mini-14	5.56mm	Gas	37.2in	6.4lbs	18.5in	5, 10, 20, 30	
AC-556	5.56mm	Gas	38.7in	6.4lbs	18.5in	20, 30	Selective fire version of Mini-14

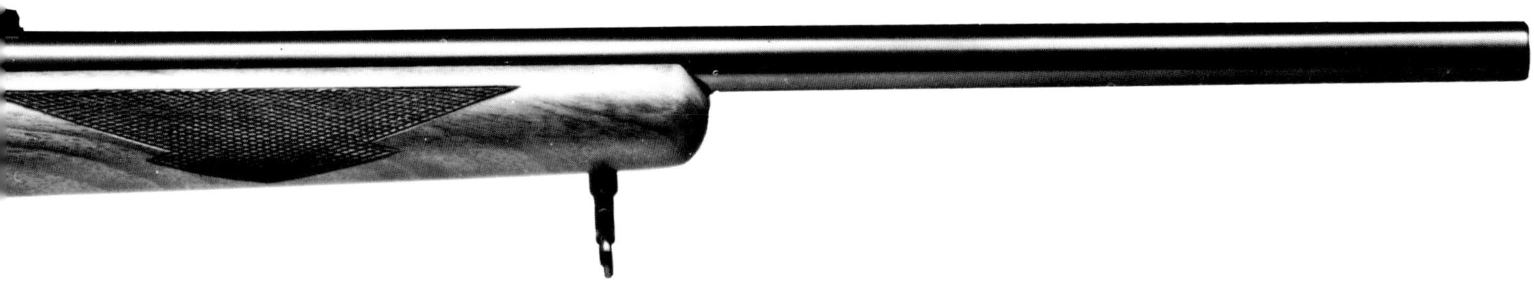

Two versions of the Ruger Model 1. Above, the No. 1 'Special Varmint' and, below, the RSI International.

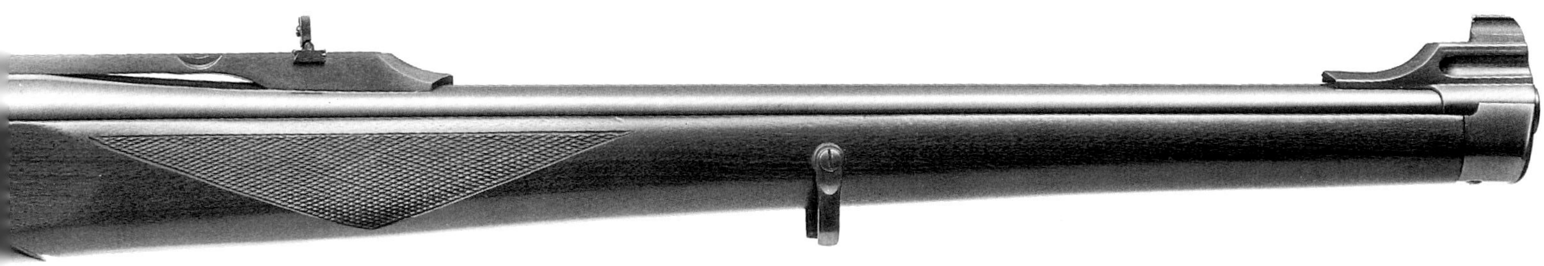

NAME	CALIBER	OPERATION	LENGTH	WEIGHT EMPTY	BARREL LENGTH	MAGAZINE CAPACITY	REMARKS
SAKO: Oy Sako AB, SF-11100 Riihimaki, Finland							
Sporter	Various	Bolt		6.75lbs	23.0in	3	Also 24in bbl
Carbine	Various	Bolt		7.5lbs	20.0in	3	Full stocked
SAVAGE: Savage Arms Corp, Westfield, MA 01085							
Model 54	.22LR	Bolt	42.0in	6.8lbs	23.0in	5	
Model 89	.22LR	Block	35.0in	5.0lbs	18.5in	S/S	Martini breech
Model 99C	.308	Lever	41.7in	6.7lbs	22.0in	5	Clip mag.
Model 99E	.308	Lever	39.7in	7.0lbs	22.0in	5	Rotary mag.
Model 110S	.308	Bolt	43.0in	8.8lbs	22.0in	5	Silhouette rifle
Model 170	.30-30	Slide	41.5in	6.8lbs	22.0in	3	
Model 340	.223	Bolt	40.0in	6.5lbs	22.0in	4	Also 24in bbl
Model 980-DL	.22LR	Blowback	40.5in	6.0lbs	20.0in	15	Tubular mag.
SIG: Schweizerische Industrie- Gesellschaft, CH-8212, Neuhausen-an-der-Rheinfalls, Switzerland							
SG510-4	7.62mm	Delay B/B	40.0in	9.4lbs	19.9in	20	Military
SG540	5.56mm	Gas	37.4in	7.2lbs	18.1in	30	Military
SG542	7.62mm	Gas	39.4in	7.8lbs	18.3in	20	Military
SG543	5.56mm	Gas	31.7in	6.6lbs	11.8in	30	Carbine version of SG540

NAME	CALIBER	OPERATION	LENGTH	WEIGHT EMPTY	BARREL LENGTH	MAGAZINE CAPACITY	REMARKS
SIG Contd.							
SG541	5.56mm	Gas	39.4in	8.8lbs	21.0in	30	Future Swiss service rifle
AMT	.308	Gas	39.0in	9.5lbs	18.8in	20	Sporter SG510-4
SINGAPORE: Chartered Industries of Singapore Pte Ltd., Jalan Boon Lay, Singapore							
SAR80	5.56mm	Gas	38.2in	8.1lbs	18.0in	20 or 30	Military
SMITH & WESSON: Smith & Wesson Inc., Springfield, MA 01101							
M1500	.270	Bolt	42.0in	7.5lbs	22.0in	5	Also 24in bbl

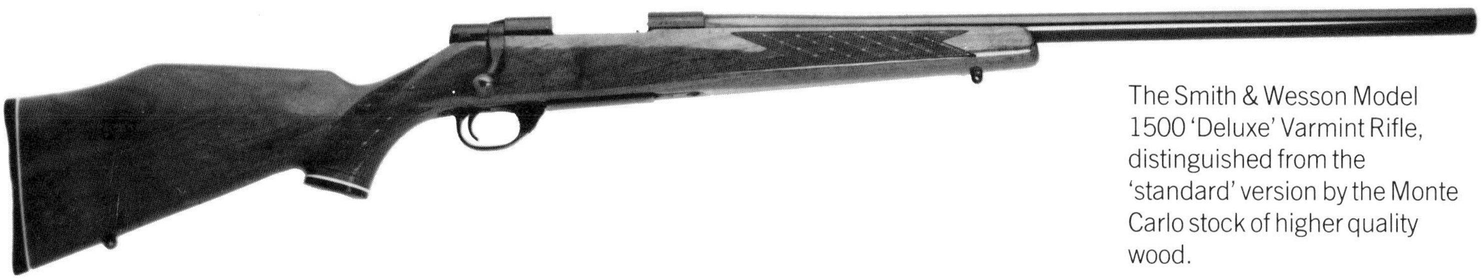

The Smith & Wesson Model 1500 'Deluxe' Varmint Rifle, distinguished from the 'standard' version by the Monte Carlo stock of higher quality wood.

NAME	CALIBER	OPERATION	LENGTH	WEIGHT EMPTY	BARREL LENGTH	MAGAZINE CAPACITY	REMARKS
SOUTH AFRICA: Armscor, Pretoria, Republic of South Africa							
R4	5.56mm	Gas	39.6in	9.5lbs	18.1in	35 or 50	Military
SPRINGFIELD: Springfield Armory Inc, Geneseo, IL							
M6	.22/.410	Break	31.5in	3.2lbs	18.0in	S/S	Survival rifle/shotgun
M14	7.62mm	Gas	44.0in	9.7lbs	22.0in	20	Commercial version of US Service rifle

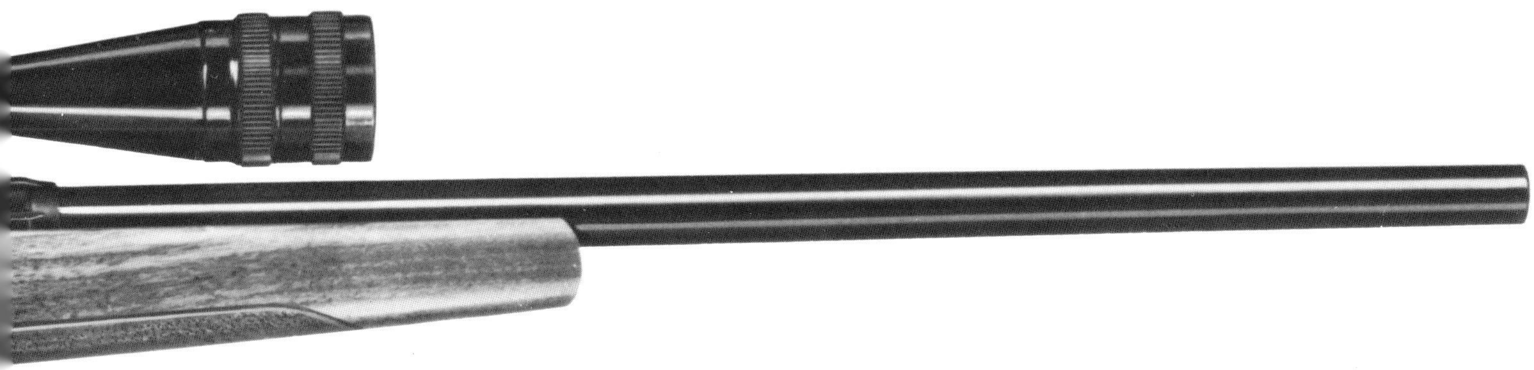

The Savage Model 110S, a heavy and accurate silhouette shooting weapon.

NAME	CALIBER	OPERATION	LENGTH	WEIGHT EMPTY	BARREL LENGTH	MAGAZINE CAPACITY	REMARKS
SQUIBMAN: Squires Bingham Mfg Co., Manila, Luzon, PI							
M16	.22LR	Blowback	38.5in	6.0lbs	16.5in	15	Replica M16A1
M20D	.22LR	Blowback	40.5in	6.0lbs	19.5in	15	
STEVENS: Savage Arms Corp., Westfield, MA 01085							
Model 110E	.30-06	Bolt	43.0ins	7.0lbs	22.0in	5	
Model 982	.22LR	Bolt	41.0in	5.8lbs	22.0in	5	
Model 987T	.22LR	Blowback	40.5in	6.0lbs	20.0in	15	Tubular mag.

The Steyr SSG-69 military sniping rifle.

NAME	CALIBER	OPERATION	LENGTH	WEIGHT EMPTY	BARREL LENGTH	MAGAZINE CAPACITY	REMARKS
STEYR: Steyr-Daimler-Puch AG, A-4400, Steyr, Austria							
AUG	5.56mm	Gas	31.0in	7.9lbs	16.0in	30	Also 13.75in, 21.5in and 24in barrels
SSG-69	7.62mm	Bolt	44.5in	10.0lbs	25.6in	5 or 10	Sniper
Model SL	Various	Bolt	38.3in	6.0lbs	20.0in	5	Full stocked
Model M	Various	Bolt	43.0in	7.5lbs	23.6in	5	Half-stock
Model S	Various	Bolt	45.0in	8.4lbs	25.6in	4	Half-stock

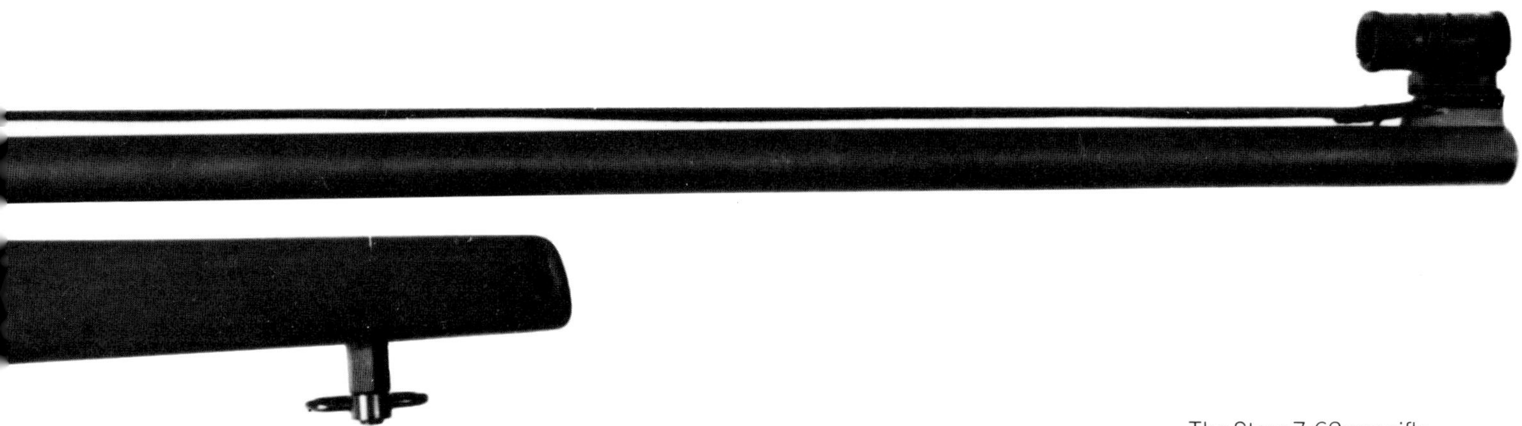

The Steyr 7.62mm rifle produced for full-bore match shooting.

NAME	CALIBER	OPERATION	LENGTH	WEIGHT EMPTY	BARREL LENGTH	MAGAZINE CAPACITY	REMARKS
STEYR Contd.							
Match-UIT	.308	Bolt	44.5in	10.9lbs	25.6in	10	Also .243 Win

Above: The Steyr AUG rifle with its alternative receiver casing which allows an image-intensifying night sight to be fitted in place of the standard optical sight incorporated in the carrying handle.

Right: Alternative barrels for the Steyr AUG, top to bottom, the light machine gun, rifle and carbine versions.

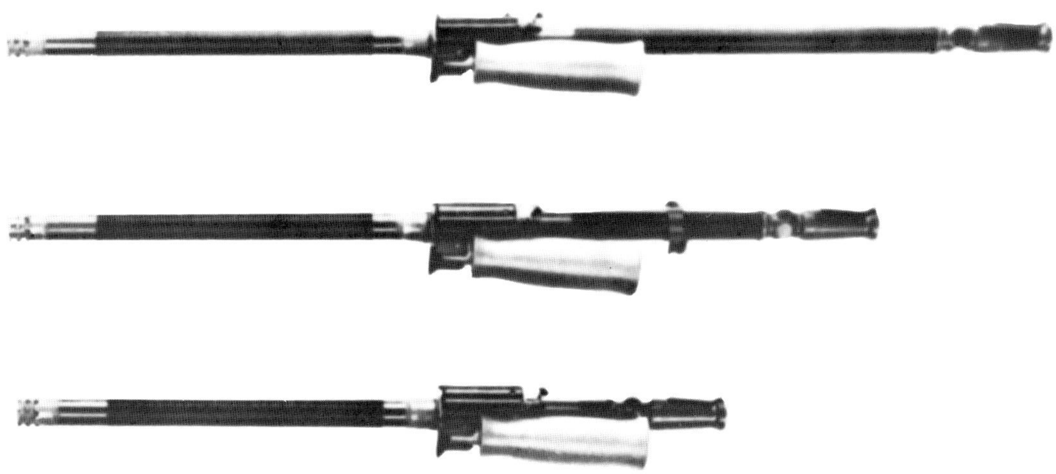

NAME	CALIBER	OPERATION	LENGTH	WEIGHT EMPTY	BARREL LENGTH	MAGAZINE CAPACITY	REMARKS
TAIWAN: Combined Service Forces, Hsing-Ho Arsenal, Taiwan							
Type 65	5.56mm	Gas	39.0in	7.0lbs	20.0in	30	Based on M16A1
UNIQUE: Manufacture d'Armes des Pyrenees Francaises, F-64700 Hendaye, France							
T-66	.22LR	Bolt	43.5in	11.5lbs	25.6in	S/S	Match rifle

NAME	CALIBER	OPERATION	LENGTH	WEIGHT EMPTY	BARREL LENGTH	MAGAZINE CAPACITY	REMARKS
U.S.A. SERVICE: Various arsenals and manufacturers							
M14	See under 'Springfield' above						
M16A2	5.56mm	Gas	39.4in	7.5lbs	20.0in	30	
M21	7.62mm	Gas	44.0in	9.7lbs	22.0in	20	M14 Sniper version
U.S.S.R.: State Arsenals							
Simonov SKS	7.62mm	Gas	40.2in	8.5lbs	20.5in	10	Military
Kalashnikov AK47	7.62mm	Gas	34.2in	9.5lbs	16.3in	30	Military
Kalashnilov AKM	7.62mm	Gas	34.5in	6.9lbs	16.3in	30	Improved AK47
Kalashnilov AK74	5.45mm	Gas	36.6in	7.9lbs	15.7in	30	Military
Kalashnikov AKR	5.45mm	Gas	26.5in		7.9in	30	Shortened AK74
UZI: Israeli Military Industries, Ramat Ha Sharon, Israel							
Uzi Carbine	9mmP	Blowback	31.2in	8.5lbs	16.1in	25	Semi-auto Uzi SMG with extended barrel. Folding butt-stock
VALMET: Valmet Corporation, Postbox 60, SF-40101, Jyvaskyla, Finland							
M76	7.62mm	Gas	36.0in	7.9lbs	16.5in	15, 20 or 30	Kalashnikov based

The Kalashnikov AK-47 as seen in a Chinese copy.

NAME	CALIBER	OPERATION	LENGTH	WEIGHT EMPTY	BARREL LENGTH	MAGAZINE CAPACITY	REMARKS
VALMET Contd.							
Model 255-470	7.62mm	Gas	27.9in	7.3lbs	16.5in	15, 20 or 30	Bullpup, under development
Model 412K	.308	Double		8.7lbs	24.0in	–	Over/under. Also in other calibers
WALTHER: Carl Walther Sportwaffenfabrik, D-7900 Ulm/Donau, West Germany							
KKMS	.22LR	Bolt	41.75in	8.61lbs	23.6in	S/S	Silhouette rifle
WA2000	.300WMag	Gas	35.6in	15.3lbs	25.6in	6	Sniper
UIT Super	.22LR	Bolt	44.8in	10.3lbs	25.5in	S/S	Match rifle
GX-1	.22LR	Bolt	44.5in	15.5lbs	25.5in	S/S	Luxury UIT
Running Boar	.22LR	Bolt	42.0in	8.4lbs	23.6in	10	Match rifle
WEATHERBY: Weatherby Inc., South Gate, CA 90280							
Fibermark	.270 WbyM	Bolt	44.37in	8.8lbs	24.0in	3 or 5	Also .240, .257, 7mm, .300 and .340 Wby Mag and .30-60
Mark V	Various	Bolt	43.3in	6.5lbs	24.0in	5	
Mark XXII	.22LR	Blowback	42.3in	6.0lbs	24.0in	5 or 10	
WHITWORTH: Whitworth Rifle Co., Manchester, England							
Express	.375H&H	Bolt	44.0in	8.5lbs	24.0in	3	
WICHITA: Wichita Arms, Wichita, KS 67211							
Classic	Various	Bolt	41.0in	8.1lbs	21.2in	S/S	
Magnum	Various	Bolt	44.8in	8.5lbs	24.0in	S/S	
Varminter	Various	Bolt	40.2in	9.0lbs	20.2in	3	
WINCHESTER: Winchester Repeating Arms Co., New Haven, CT 06504							
Model 70	Various	Bolt	42.5in	7.5lbs	22.0in	3	
Model 94	.30-30	Lever	37.8in	6.5lbs	20.0in	6	Tubular mag.
Model 9422XTR	.22RF	Lever	37.2in	6.5lbs	20.5in	15-21	Tubular mag.
ZASTAVA: Zavodi Crvena Zastava, Kragujevac, Yugoslavia							
M59/66	7.62mm	Gas	44.1in	9.0lbs	19.7in	10	Simonov based
M70B1	7.62mm	Gas	35.4in	8.2lbs	16.3in	30	Kalashnikov based
M76	7.62mm	Gas	44.7in	9.2lbs	21.6in	10	Sniper
M77B1	7.62mm	Gas	38.9in	10.5lbs	19.7in	20	Export model
M80	5.56mm	Gas	39.0in	7.7lbs	18.1in	30	Export model

INDEX

Page numbers in italics refer to illustrations

ACKNOWLEDGMENTS

The author and publishers would like to thank the various manufacturers who supplied pictures of their products for reproduction in this book. The publishers would also like to thank the London Armoury for their help in preparing the photograph on pages 2-3 which was taken by John Freeman. Thanks also to Ron Watson who compiled the index and Design 23 who designed this book.